A Developer's Guide to Cloud Apps Using Microsoft Azure

Migrate and modernize your cloud-native applications with containers on Azure using real-world case studies

Hamida Rebai Trabelsi

BIRMINGHAM—MUMBAI

A Developer's Guide to Cloud Apps Using Microsoft Azure

Group Product Manager: Rahul Nair
Publishing Product Manager: Surbhi Suman
Senior Editor: Divya Vijayan / Romy Dias
Technical Editor: Arjun Varma
Copy Editor: Safis Editing
Project Coordinator: Ashwin Kharwa
Proofreader: Safis Editing
Indexer: Manju Arasan
Production Designer: Ponraj Dhandapani
Marketing Coordinator: Nimisha Dua

First published: January 2023

Production reference: 1190123

Published by Packt Publishing Ltd.
Livery Place
35 Livery Street
Birmingham
B3 2PB, UK.

ISBN 978-1-80461-430-3

www.packt.com

To my mother, Rafika Boukari, and to the memory of my father, Mongi Rebai, for their sacrifices, their blessings, and the constant support that they have been for me. To my husband, Mohamed Trabelsi, for supporting me during my writing journey. To my son, Rayen Trabelsi, and my baby daughter, Eya Trabelsi, for their smiles every day that keep me always motivated.

Foreword

While I don't have a crystal ball, perhaps in a few years we will look back at this digital age and consider it to be an inflection point in how we host, build, and manage applications. Cloud services have already transformed how developers can build amazing applications and services, and this trend is only starting; there is so much more to come. As I am writing this foreword, we can see that AI will become the next tool in the developer's toolbox, making it easier to transform amazing ideas into reality.

The adoption of cloud-native services, containers, and orchestration platforms as key building blocks in application architectures may appear to you like new(ish) ideas, but ultimately, I believe that this book will show you how they can enable you to deliver more value and focus on what counts the most. Capabilities that not too long ago were reserved for mega-scale scenarios have become available and accessible to all types and sizes of applications.

I have had the opportunity to collaborate with Hamida over the last few years on a number of initiatives in my role at Microsoft and witnessed first hand how she has been a change driver of this transformation, notably in terms of guiding individuals and organizations in leveraging the power of the cloud, but perhaps more importantly, in adopting a growth mindset, fostering a learning culture, and embracing change and curiosity.

I truly think that we are on a lifelong journey of change, and we should not be daunted by this, but rather embrace it, and you may be surprised where this will take you.

This book will provide you with meaningful and practical scenarios and examples that I think will resonate with you, regardless of where you are in your journey as a cloud practitioner, and give you new perspectives on how you can build highly resilient and scalable applications.

Throughout the chapters of this book, you will go through several technical scenarios where Hamida provides guided steps on how this can be implemented. I will leave you with this very small proposal: make sure you set some time aside for hands-on experimentation. It has become very simple to leverage a number of free options for practice and learning, such as Microsoft Learn sandboxes, and, like Hamida, I have found them extremely useful to fuel my curiosity and creativity.

I wish you success in your journey.

Marc-Andre Laniel

Senior Program Manager, Enterprise Skilling @Microsoft

Contributors

About the author

Hamida Rebai Trabelsi has been working in the computing domain for over 12 years. She started her professional career in Tunisia working for MNCs as a software developer, then served as a .NET consultant at CGI, Canada, and currently, she is a senior advisor and information and solution integration architect at Revenu Québec, Canada. She has been awarded Most Valuable Professional in Developer Technologies and Microsoft DevHero by Microsoft and holds several Azure certifications. Besides being a Microsoft Certified Trainer and a member of .NET Foundation, Hamida is a blogger, an international speaker, and one of the finalists in the Women in IT Award Canada 2019.

This is my first book, and I have been overwhelmed by the support I have received from the Packt team, who encouraged me during the process. I feel proud that I was able to finish this book and share my experience as a developer starting with Azure. To the memory of my father, Mongi Rebai, I offer this book to you and I hoped that you were with me sharing this new adventure in my professional career. Thank you to my mum, Rafika Boukari, for her sacrifices and support. Thank you to my husband, Mohamed Trabelsi, for believing in me. Every day, I wake up and see my son, Rayen Trabelsi, and my baby girl, Eya Trabelsi, and I feel more motivated, so thank you for this energy.

About the reviewers

Ryan Mangan is a cloud and an end user computing technologist focusing on application migration and modernization. He's an international speaker on a wide range of topics, an author, and has helped customers and technical communities over the past decade. Ryan is a chartered fellow of the British Computer Society, a Microsoft **Most Valuable Professional** (**MVP**), as well as a VMware vExpert, and **Very Important Parallels Professional** (**VIPP**).

Stefano Demiliani is a Microsoft MVP on Azure and business applications, a **Microsoft Certified Solution Developer** (**MCSD**), an Azure Certified Architect, and an expert in other Microsoft-related technologies. His main activity is architecting and developing enterprise solutions based on the entire stack of Microsoft technologies (mainly focused on ERP and serverless architectures). He has worked with Packt Publishing on many IT books related to Azure cloud applications and Dynamics 365 Business Central, and is a frequent speaker at conferences around Europe. You can reach him on Twitter (@demiliani) or LinkedIn.

Table of Contents

Part 1 – Migrating Applications to Azure

1

2

3

Migrating Your Existing Applications to a Modern Environment   31

4

Exploring the Use Cases and Application Architecture   61

Part 2 – Building Cloud-Oriented Applications Using Patterns and Technologies in Azure

5

6

7

Part 3 – PaaS versus CaaS to Deploy Containers in Azure

8

9

Understanding Container Orchestration   163

10

Setting Up a Kubernetes Cluster on AKS   175

Part 4 – Ensuring Continuous Integration and Continuous Deployment on Azure

11

12

Preface

Azure is a public Cloud computing from Microsoft provider. It provides a range of Cloud services including compute, network, storage and analytics. Users are able to select from them to develop and scale new or existing application in the Cloud.

Further development of applications, processes, and infrastructure is critical to successfully modernizing your organization and achieving your strategic goals. To increase business value, enterprises are adopting new application architectures that can deliver new capabilities quickly and flexibly. Agile business processes optimize operational efficiency and accelerate response to changing market conditions. Essentially, cloud-based infrastructure allows companies to control costs and, importantly, benefit from new innovations. Containers are one of the key technologies for modernizing and optimizing IT.

The goal of this book is to modernize your application by adopting containers and processes. We'll learn which Azure services to choose to run your application and how select the suitable service based on your needs. There are many options in Azure to run your applications. These range from VMs to Container to Cloud Services and we will explore the foundational knowledge of the Microsoft Azure landscape that will help you as you move forward to pick the right services in Azure for your applications.

Who this book is for

This book is essentially intended for Cloud developers, software architects, system administrators, developers, and computer science students looking to understand the new role of software architect or developer in the cloud world.

This book is also for Professionals looking to enhance their cloud and cloud-native programming concepts will also find this book useful and looking to explore the possibilities on Azure. But they should have a strong background in C#, ASP.NET Core, Visual Studio (any recent version) and a basic knowledge of cloud computing will be helpful when using this book.

What this book covers

Chapter 1, An Introduction to the Cloud-Native App Lifecycle, starts by introducing a basic concept of building and deploying Cloud-based apps. Cloud-native is the core of application innovation and modernization. It compares monolithic and microservices and their architectures. It also briefly discusses the application lifecycle dans design and serverless applications that help to build and run scalable applications. It focusses also on the 12 Factor Application design methodology.

Chapter 2, Beginning Your Application Migration, covers Cloud Adoption Framework that is a guidance and best practices to adopt the Cloud. It also briefly discusses the difference between Cloud Migration and Cloud Adoption and the different steps to consider to prepare the migration. Cloud Adoption Strategy Assessment and Cloud rationalization are discussed. It evaluates different migration considerations. Azure presents different hosting options to migrate existing or new application using Azure Virtual machines as Infrastructure as a service (IaaS), and different Azure service as platform as service (PaaS).

Chapter 3, Migrating Your Existing Applications to a Modern Environment, begins with the benefits of moving legacy apps to the cloud and modernizing the existing legacy applications using a set of new technologies and approaches. It covers the migration of an ASP.NET Web solution to different environments: Azure VM using Microsoft migration tools and services or Azure App Service or Windows container. It covers also the database migration to Azure using different tools and services including the assessment phase.

Chapter 4, Exploring the Use Cases and Application Architecture, presents the solution reference and its architecture that will be used in the book to understand the different services to be used in Azure. This chapter will define the difference between monolithic architecture and microservices architecture. To build large-scale applications, we need to consider the different challenges and the solutions for distributed data management.

Chapter 5, Learning Cloud Patterns and Technologies, covers the different Cloud Design Pattern and technologies to build reliable, scalable and secure Cloud-based applications.

Chapter 6, Setting Up an Environment to Build and Deploy Cloud-Based Applications, covers the environment preparation to build Cloud-Native application. The book will define the prerequisites. It presents the different tools, framework, and technologies in Windows and Linux.

Chapter 7, Using Azure App Service to Deploy Your First Application, covers the creating and the deploying of an Azure App Service resource using different operating systems and platforms. This chapter will help you to select the right App Service plan and exploring a set of features like App Service deployment slots and scaling options. It discusses the deployment of a Cloud-based application to App Service.

Chapter 8, Building a Containerized App Using Docker and Azure Container Registry, starts by discussing the development process for Docker-based applications. It starts by defining the basic concepts related to Docker container, images dans Dockerfile implementation. This chapter presents Azure Service to deploy containers in the Cloud. It discusses the creating of Azure Container Registry and Azure Container Instances..

Chapter 9, Understanding Container Orchestration, covers the container orchestration. This chapter start by comparing between Docker containers and orchestrators and Kubernetes cluster architecture. This chapter will present Azure Kubernetes Services and Azure Container Apps. For multi-container applications, we will deploy microservices using Docker Desktop and Kubernetes.

Chapter 10, Setting Up a Kubernetes Cluster on AKS, covers the creating of Kubernetes Cluster on Azure Kubernetes Service. This chapter will discuss the deployment of an application to the cluster. Azure DevOps Starter will be detailed to deploy a Kubernetes Cluster. To deploy an application deployed in Azure Kubernetes Services, we will present Bridge to Kubernetes extension in Visual Studio.

Chapter 11, Introduction to Azure DevOps and GitHub, starts by defining DevOps and the basic concepts. In this chapter, we will discuss about DevOps tools like Azure DevOps and GitHub.

Chapter 12, Creating a Development Pipeline in Azure DevOps, covers the setting up of Azure DevOps environment. In this book, we will create a build pipeline and release pipeline in Azure Pipeline and we will create a continuous integration and continuous deployment pipeline for GitHub repo using Azure DevOps Starter.

To get the most out of this book

The most of examples presenting the solution reference are presented using .NET 6, basic knowledge in C# and .NET technology is essential to understand described topics.

Software/hardware covered in the book	Operating system requirements
.NET 6 and .NET 7	Windows, macOS, or Linux
Docker Desktop	Windows, macOS, or Linux
Visual Studio 2022 (community edition)	Windows or macOS
Visual Studio Code	Windows, macOS, or Linux
Service fabric runtime	Window or Linux

We assume that you are able to install Visual Studio Code as an IDE or Visual Studio 2022, the community edition is enough to run the examples if you don't have access to the commercial license.

If you are using the digital version of this book, we advise you to type the code yourself or access the code from the book's GitHub repository (a link is available in the next section). Doing so will help you avoid any potential errors related to the copying and pasting of code.

All examples are used to explain the use of every Azure Service.

Download the example code files

You can download the example code files for this book from GitHub at `https://github.com/PacktPublishing/A-Developer-s-Guide-to-Cloud-Apps-Using-Microsoft-Azure`. If there's an update to the code, it will be updated in the GitHub repository.

We also have other code bundles from our rich catalog of books and videos available at `https://github.com/PacktPublishing/`. Check them out!

Download the color images

We also provide a PDF file that has color images of the screenshots and diagrams used in this book. You can download it here: `https://packt.link/63FuQ`.

Conventions used

There are a number of text conventions used throughout this book.

`Code in text`: Indicates code words in text, database table names, folder names, filenames, file extensions, pathnames, dummy URLs, user input, and Twitter handles. Here is an example: "You can add the `--deployment-local-git` parameter at the end if you use Git to deploy your application."

A block of code is set as follows:

```
apiVersion: apps/v1
kind: Deployment
metadata:
  name: booking-frontend-deployment
spec:
```

When we wish to draw your attention to a particular part of a code block, the relevant lines or items are set in bold:

```
apiVersion: v1
kind: Service
metadata:
  name: booking-frontend-service
```

Any command-line input or output is written as follows:

```
$ az appservice plan create --name myserviceplan --resource-
group packtrg --sku D1 --is-linux
```

Bold: Indicates a new term, an important word, or words that you see onscreen. For instance, words in menus or dialog boxes appear in **bold**. Here is an example: "Open your solution, right-click on the application to containerize, and then select **Add** and then **Docker Support...**."

> Tips or important notes
> Appear like this.

Get in touch

Feedback from our readers is always welcome.

General feedback: If you have questions about any aspect of this book, email us at customercare@ packtpub.com and mention the book title in the subject of your message.

Errata: Although we have taken every care to ensure the accuracy of our content, mistakes do happen. If you have found a mistake in this book, we would be grateful if you would report this to us. Please visit www.packtpub.com/support/errata and fill in the form.

Piracy: If you come across any illegal copies of our works in any form on the internet, we would be grateful if you would provide us with the location address or website name. Please contact us at copyright@packt.com with a link to the material.

If you are interested in becoming an author: If there is a topic that you have expertise in and you are interested in either writing or contributing to a book, please visit authors.packtpub.com.

Share your thoughts

Once you've read *A Developer's Guide to Cloud Apps Using Microsoft Azure*, we'd love to hear your thoughts! Scan the QR code below to go straight to the Amazon review page for this book and share your feedback.

https://packt.link/r/1804614300

Your review is important to us and the tech community and will help us make sure we're delivering excellent quality content.

Download a free PDF copy of this book

Thanks for purchasing this book!

Do you like to read on the go but are unable to carry your print books everywhere?

Is your eBook purchase not compatible with the device of your choice?

Don't worry, now with every Packt book you get a DRM-free PDF version of that book at no cost.

Read anywhere, any place, on any device. Search, copy, and paste code from your favorite technical books directly into your application.

The perks don't stop there, you can get exclusive access to discounts, newsletters, and great free content in your inbox daily

Follow these simple steps to get the benefits:

1. Scan the QR code or visit the link below

https://packt.link/free-ebook/9781804614303

2. Submit your proof of purchase
3. That's it! We'll send your free PDF and other benefits to your email directly

Part 1 – Migrating Applications to Azure

In this part of the book, we will present the cloud migration journey and the benefits of moving apps to the cloud.

This part comprises the following chapters:

- *Chapter 1, An Introduction to the Cloud-Native App Lifecycle*
- *Chapter 2, Beginning Your Application Migration*
- *Chapter 3, Migrating Your Existing Applications to a Modern Environment*
- *Chapter 4, Exploring the Use Cases and Application Architecture*

1

An Introduction to the Cloud-Native App Lifecycle

This first chapter is about introducing the basic concepts of cloud-native development and the lifecycle involved. You will learn the basic concepts behind building and deploying applications on any cloud platform, including adopting a **microservices** architecture, **containerization**, and **orchestration**.

To enable developers to build applications with more flexibility and more portability compared to applications hosted on traditional servers or **virtual machines** (**VMs**), we will learn how to use containers and serverless architecture.

To accelerate the product development process and improve the quality of our apps, we will follow the Twelve-Factor Application design principles and methodology. As projects grow, code bases also become more complex, and it is strongly recommended that you test new versions of software.

In this chapter, we're going to cover the following main topics:

- An introduction to cloud-native applications
- Application design
- Application lifecycles
- The Twelve-Factor Application design methodology
- Serverless applications of cloud-native applications

An introduction to cloud-native applications

In the continuously competitive market of digital companies, the main issue is IT agility.

However, with technology evolving every day, companies are struggling to catch up with digital transformation, adopt new trends such as the cloud, **artificial intelligence** (**AI**), mobility, or the **Internet of Things** (**IoT**), and change their business models in order to be able to adapt to the new reality of the market.

This delay is caused by the interdependencies of services related to the lifecycle of an IT project, where changes to the source code of a classic monolithic client-server application can destabilize systems in the maintenance phase.

For this reason, companies are looking to adopt more optimized, scalable architectures to ensure resiliency and continuous availability, while minimizing resource consumption costs and implementing cloud-native applications.

This change should be made quickly, especially for start-ups that want to build rigid infrastructures with minimal costs.

A *cloud-native* approach includes all the concepts of building and deploying scalable applications in modern, dynamic environments on any cloud platform, be it a public, private, or hybrid cloud. Organizations use cloud-native technologies to build and run scalable applications.

Today, application environments are modern, automated, and dynamic because we can publish any application and store data in a public, private, or hybrid cloud. We can use technologies such as containers, service fabric, and immutable infrastructure, and patterns such as microservices, declarative APIs, and event buses. These techniques help us implement loosely coupled solutions that are easy to maintain, resilient, and observable.

The robust automation of environments, as in the case of *infrastructure as code*, allows engineers to make major changes frequently without worrying about the impact and with minimal effort.

A cloud-native application has specific features that present the pillars of cloud-native architecture and a design pattern, used to build a new application from scratch that can be deployed in a cloud environment.

Microservices are a part of cloud-native application architecture and run as a bundle independently of each other on a containerized and orchestrated platform. They connect and communicate through APIs.

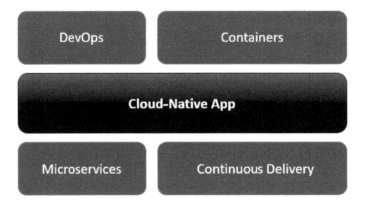

Figure 1.1 – The pillars of cloud-native architecture

Applications based on cloud-native application architecture are reliable and provide scalability and performance to meet the recurring goal of achieving a fast time to market, if they are well designed.

Application design

One of the most important steps in the lifecycle of an application is the design and architecture of the app. This is the most critical aspect prior to starting the implementation, deployment, continuous delivery, and maintenance of an application. The evolution in technologies and patterns always influences the design of applications as we seek to ensure performance and security.

What happens if the application stops working, or crashes for no reason or due to a lack of resources? How are you going to debug it if the error isn't clear, or if the logs aren't really good enough?

If you need to ask these questions, then you are on the wrong path – you are not working in a cloud-native application context.

The design of cloud-oriented applications has the objective of taking advantage of the benefits of the cloud. Even the software and services that manage these applications will be deployed in the cloud.

Cloud-native applications are typically microservices embedded in containers running on cloud computing infrastructure. Cloud-native applications use a microservice architecture. This architecture ensures the allocation of resources to each service used by the application more efficiently than the old approaches, such as monolithic applications. This makes an application flexible and adaptable to a cloud-oriented architecture.

Monoliths versus microservices

It is really important to understand the difference between the traditional monolithic approach and the microservices approach before defining the concept of microservices.

To scale a monolithic application, we have to clone the entire application on multiple servers or VMs. But for a microservices application, scaling is done by deploying every service in multiple instances across servers or VMs.

In a microservices approach, every application is composed of a collection of services that are related to specific functionalities. Every service can be developed, tested, deployed, versioned, and scaled. Monolithic applications are simple to use and easy to develop, deploy, and test, but they have limitations in size and complexity. Despite the simplicity of horizontal scaling, where we run multiple application copies behind a load balancer, it is difficult to do in the case of multiple modules that have conflicting resource requirements.

Microservices are very similar to beehives. Within a hive, thousands of bees coexist and help each other for a single common goal – the survival of the colony. The queen, the workers, and the drones – each one has their peculiarities and must, therefore, assume distinct tasks.

Monolithic and microservices architectures

Microservices are regularly discussed now in articles, on blogs, on social media, and even at conference presentations.

How do we use microservices to tackle complexity? In recent years, software architecture has evolved rapidly, from spaghetti-oriented architected where everything was a big jumble, to lasagna-oriented architecture where we can see the layers of architecture, to ravioli-oriented architecture where we talk about microservices. In this latter architecture, we split the application into sets of services instead of building a monolithic application. Maybe we will see *pizza-oriented architecture* next, where we use a serverless approach. Let's now take a look at the layered architecture pattern and compare it to microservices.

The layered architecture pattern is the most common architecture pattern and is otherwise known as the **n-tier architecture** pattern. For distributed n-tier client/server applications, when taking a monolithic approach, you start with a **hexagonal modular architecture**, where you separate the domain model and the adapters (the devices used for inputs and outputs).

A monolithic application is composed of several layers, including different types of components or layers.

In this classic example, illustrated in *Figure 1.2*, we have four layers, from the user interface to the database where we store our data:

- **Presentation layer**: This presents the user interface layer; it can be a web or mobile or desktop application.

- **Services layer**: This is a set of standards, techniques, and methods. An application is split into services based on functionality and domain. This layer is responsible for handling HTTP requests and responding with either HTML or JSON/XML, as in the case of API services.

- **Business logic layer**: This holds the application's business logic, the heart of our application. This entails custom business rules, such as operation definitions, constraints, and algorithms, that manage the exchange of information between the database layer and the presentation layer.

- **Database access layer**: This is an abstraction of the logical data model. The modification of the logical data model is done in the business layer, but we can perform even more complex data manipulations from multiple sources and send them back to the business layer. This layer will ensure access to the database.

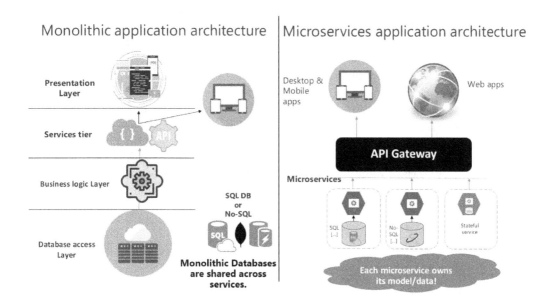

Figure 1.2 – N-tier architecture pattern for monolithic and microservices architectures

The architecture presented here is logically modular, but the application is packaged and deployed in a single package as a monolith.

Let's explore the different elements of the monolithic approach:

- Monolithic applications are easy to set up because they involve a single complete package. They include all the components, such as the GUI, business layer, data, and related necessary services encapsulated in a single package.

- This single package is developed in sequential order. At the start of each project, the designers work on the design of the product as well as the necessary documents to meet the needs of the client user. Then, the developers implement code and send it to the quality assurance department in order to carry out the necessary tests.

- The team of testers runs different types of tests, including integration tests, interface tests, and even performance tests, to identify errors and evaluate the performance of the cloud-native application.

- If they detect errors, code is sent back to the developers so that it can be debugged and corrected.

- Once the code passes all the tests, it is deployed in a test production environment similar to the final environment and then deployed in a real environment.

- If you want to modify the code, add a new feature, or even remove an old feature, you have to start the whole process again. If several teams are working on the same project, taking into account the changes in the teams as developers come and go, coordinating code changes is a challenge and will take a lot of time. Moreover, the deployment of a software project always requires a specific infrastructure configuration as well as the implementation of an extended functional test mechanism. Therefore, in a complex project with dynamic requirements, the whole process is inefficient and time-consuming. Microservices architecture can help to solve most of these challenges.

For microservices, the idea is very simple – we divide our application into a set of smaller interconnected services instead of creating a single monolithic application. This is an architectural design model based on the architectural principles of **domain-driven design** and **DevOps** – that is, if we have several domains in our application that can interact independently.

Each microservice represents a small application with its own hexagonal architecture. This application is composed of business logic and data.

Some microservices may implement a user interface, such as a web page or mobile application, while others may expose or consume a **Representational State Transfer** (**REST**), **Application Programming Interface** (**API**), or even a **Remote Procedure Call** (**RPC**). Services may be using a message-based system or simply consuming APIs provided by other services.

The services are independent, making it easy for developers to work independently on them without affecting the entire app. Developers also have the freedom to use different languages in different parts of the code simultaneously, via a central repository that acts as a version control system and updates specific features without disrupting the software or causing application downtime.

Developers can use a central container orchestrator to improve performance by managing automatic scheduling and the allocation of resources on demand, by using scaling features. Finally, microservices are adopted as enablers of DevOps and **Continuous Integration** (**CI**)/**Continuous Delivery** (**CD**), allowing them to be updated and deployed faster and without issues.

Microservices are deployed independently, each service having its own database, as shown in the previous diagram.

Microservices allow us to scale and deploy parts of an application independently. They offer great distributed software challenges, but with these benefits, microservices are not a universal solution for every app in the cloud, as they are intended mainly for large, scalable, and long-term distributed applications. Do not underestimate the complexity involved in implementation and testing.

We have discussed the application design and have explored the difference between monolithic and microservices approaches. Let's now move on to application lifecycles.

Application lifecycles

The term *application lifecycle* refers to the cyclical software development process, which includes planning and monitoring, development, testing, deployment, operation, monitoring, collaboration, and communication.

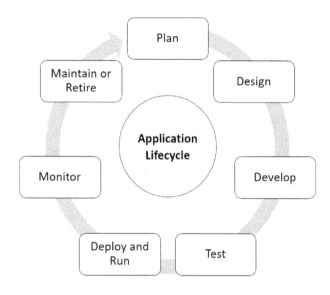

Figure 1.3 – Application lifecycle

Application Lifecycle Management (**ALM**) entails the use of a set of tools, teams, and processes to manage the lifecycle of an application, from requirements management, project management, design and software architecture, development, unit and integration testing, maintenance, update requirement management, CI, delivery integration, deployment, and release management to the end of life.

ALM consists of the following five stages.

Stage 1 – application governance

Application governance is the initial stage of decision-making and includes requirements management. During this stage, the team begins to define the functions and functionalities of the application that are required to achieve the objectives defined by the client. This involves designing the concept of an app based on these user requirements.

Stage 2 – development

This is the most important stage in the application lifecycle because this stage determines the creation of the application. The developers take the functionalities planned in the previous step and prepare a development plan to achieve them. In most cases, these functionalities will be broken down into chunks and then assigned to the appropriate teams to develop a schedule for the release of each phase. After creating the application, the teams then start implementing code and integrating it according to the plan.

Stage 3 – quality assurance – software testing

Once the application has been implemented in line with the requirements, the next stage is the testing phase to ensure that the application actually meets all the requirements, works without errors, and provides an appropriate user experience. Test scenarios and environments are prepared and application performance testing is performed. The testers provide feedback at the end and publish reports on errors encountered, including unconfirmed ones and even bugs, and the development team updates the product based on this feedback.

Stage 4 – deployment

This stage begins when the product is ready to be deployed to production for end users. This can be done via several methods, depending on the needs of the customers. A continuous deployment strategy can be put in place to facilitate the automation of this process.

Stage 5 – operations and maintenance

The ALM process does not stop at the point of product deployment to users – it continues with the ongoing operation and maintenance of the product. To confirm that the software is meeting the business objectives, in-use performance monitoring should be put in place to prevent overloads or service downtime issues. This also allows the team to find and resolve any problems encountered, along with providing updates and improvements.

The final phase of this stage involves the withdrawal of the product according to criteria defined in advance by the team. This details the reason for the decision to withdraw the software and move to a new version or a new product.

Now we have discussed the application lifecycle model, let's now move on to the Twelve-Factor Application design methodology.

The Twelve-Factor Application design methodology

Nowadays, software is usually provided as a service, whether in the form of **web applications** or **software as a service** (**SaaS**). The Twelve-Factor App is an influential software application design model for designing a scalable application architecture.

> **Note**
>
> The Twelve-Factor App was published in 2011 by *Adam Wiggins* and provides a set of principles to follow in order to create code that can be released reliably, updated quickly, and maintained consistently.

The Twelve-Factor App methodology page can be found at `https://12factor.net/`. The following is a summary of the principles:

- **Code base** – "One code base tracked in revision control; many deploys":

  Each application must have its own code base (or repository); multiple applications should not share the same code. A code base is a repository of versioned code. However, we must avoid creating multiple code bases for different versions – version management must be managed by a repository tool such as Git. It is recommended that for all deployment environments, there should be only one repository, not multiple.

- **Dependencies** – "Explicitly declare and isolate dependencies":

  Every application has dependencies with other packages. Most of the dependencies require the use of external dependencies, and the objective is to deploy the application with its dependencies because they form a *whole*, a kind of bundle.

  Therefore, you have to declare the dependencies explicitly and precisely before creation, and then isolate these dependencies at runtime.

  This is enabled through **NuGet** tools in **.NET Framework** or npm for JavaScript. These tools define their dependencies inside the manifests, including very specific versions, and have the role of then ensuring that the dependencies are running correctly.

- **Config** – "Store config in the environment":

  The idea is to separate the code from the configuration of the application itself. This configuration can be placed in environment variables, which will be injected at runtime, and the configuration and the code will thus be in separate files. However, sensitive data such as credentials and keys should not appear visibly in the code, nor even in the configuration file, for security reasons. Good examples of external configuration files are the `appsettings.json` file in .NET projects, a Kubernetes manifest file, and a `docker-compose.yml` file.

- **Backing services** – "Treat backing services as attached resources":

  A support service is any service your application needs for its functionality, such as databases, mail servers, mail systems, caching systems, and even services running business functionality and security.

- **Build, release, and run** – "Strictly separate build and run stages":

  The build process focuses on building everything that is needed for your application, the release stage combines the output of the previous stage (the build stage) with the configuration values (both environmental and application-specific), and the run stage uses tools such as containers and processes to launch the application, meaning that it needs a specific environment to run the application that's distinct from the environments used in the build and release stages.

- **Processes** – "Execute the app as one (or more) stateless process(es)" :

  We are talking here about the process state. The application can work as a collection of stateless processes, implying that no trace of the state of another process (such as the state of the session) will be saved. Equally, the workflow and instances can be added and removed to handle a particular workload at a given time. A stateless process makes scaling easier. In conclusion, each process is independent of the others, which prevents surprises.

- **Port binding** – "Export services through port binding":

  An application is identified in the network by a port number or a domain name known to the **Domain Name System (DNS)**. The idea behind the principle of port binding is that the use of ports in the network is very efficient – for example, port 80 is used for web servers running under HTTP, port 443 is the default port number for HTTPS, port 22 is for SSH, port 3306 is the default port for MySQL, and port 27017 is the default port for MongoDB.

- **Concurrency** – "Scale-out through the process model":

  The concurrency factor states that applications should be able to scale up or down elastically, depending on their workload.

- **Disposability** – "Maximize robustness with fast startup and graceful shutdown.":

  The availability principle states that applications should start and stop properly without slowness or errors. This means that users can access the application without experiencing service downtime issues.

  In the event of a shutdown, it is recommended to ensure that all database connections and other network resources are properly terminated and that all shutdown activity is logged.

- **Dev/prod parity** – "Keep development, staging, and production as similar as possible":

  The dev/prod parity factor focuses on the importance of keeping the development, simulation, acceptance, and production environments as similar as possible. But why? Because it is important to identify potential bugs and errors during development and testing before an application is released to production, all deployment environments are similar but independent.

- **Logs** – "Treat logs as event streams":

  The logs factor highlights the importance of ensuring that your application doesn't itself manage the routing, storage, or analysis of its output stream (i.e., logs).

  For example, one consumer might be interested in error data, but another consumer is interested in request/response data. Yet another consumer is interested in storing all log data for event archiving. This means that logs should be treated as a stream of log events. If we remove an application, the log data persists long after.

- **Admin processes** – "Run admin/management tasks as one-off processes":

  This factor recommends not setting up one-time administration or management tasks in the application.

The examples given on `https://12factor.net/` are for migrating databases and running one-time scripts to perform cleanup.

Now that we have discussed the Twelve-Factor App design methodology, let's now move on to serverless applications.

Serverless applications

In the world of cloud computing, in order to better understand public cloud services such as Microsoft Azure, it is necessary to understand the shared responsibility model and distinguish between what will be managed by the cloud provider and the tasks that are your responsibility to manage.

Workload responsibilities vary depending on the workload. These workloads can be hosted as **Software as a Service (SaaS)**, **Platform as a Service (PaaS)**, **Infrastructure as a Service (IaaS)**, or in an on-premises data center. What is interesting about cloud providers is that they provide the infrastructure required to run applications for the users. Cloud providers support the execution of the servers deployed to the dynamic management of the resources of the machine. These machines can be scaled according to the runtime load.

The user focuses only on the development and deployment of applications.

Serverless computing, also known as **Function as a Service** (**FaaS**), is a cloud-native development model that allows developers to build and run applications without having to manage servers. Serverless computing allows developers to only write the code, while the backend infrastructure is managed by the cloud provider. Developers can write multiple functions in order to implement business logic in an application, and then all these functions can be easily integrated to communicate with each other. Applications using this pattern are said to be using **serverless architecture**.

Microservices and serverless are two major concepts in cloud computing today.

Serverless architecture is a very commonly implemented aspect of microservices architecture. In microservices architecture, the application is broken down into small independent pieces and each one has its own task to fulfill. Deployment and management of microservices are widely used in a serverless model.

In a serverless model, application code is executed on demand in response to pre-configured triggers by the application developer. However, the benefits of building applications from microservices are perhaps most apparent when the application is hosted in the cloud using serverless architecture.

For most use cases, code is executed in stateless containers. Code execution can be triggered by different events, such as sending HTTP requests, database events, and queuing services. Meanwhile, in the application, we can trigger monitoring alerts, or even initiate file downloads and scheduled events (as in the case of **cron jobs**).

FaaS is a subset of serverless computing, focusing on event-driven triggers where code is executed in response to events or requests.

The use of serverless computing improves developer productivity and reduces the time required to release and deliver new applications to the market. However, along with the benefits of serverless computing, challenges also present themselves in the difficulty of monitoring and maintaining serverless applications. Consideration should also be given to the security issues around serverless architecture, such as the need to analyze short-lived functions in order to scan for vulnerabilities and prevent code injection.

The challenges of cloud-native applications

Cloud-native applications take advantage of the cloud operating model, the benefits of which we discussed previously. However, as well as benefits, there are also challenges with cloud-native development that every organization should consider before beginning their move to it.

Although the theory behind the development of cloud-oriented applications seems clear and simple enough, problems remain at the level of implementation, especially if an enterprise has longstanding legacy applications.

Let's take a look at some of the most common challenges faced by enterprises in their cloud-native journeys.

The challenges of service discovery and CI/CD pipelines for microservices applications

If we have several microservices that communicate with each other, these microservices run in different instances. The number of service instances and their locations change dynamically. The number of service instances and their locations change dynamically. The service discovery mechanism helps us to locate each instance.

CI encourages continuous code merging and testing, leading to the early detection of bugs. Other benefits include less time wasted dealing with merge issues and faster feedback to the development team.

CD is an extension of CI. It is a semi-manual process that allows developers to deploy all changes to their customers with a simple click of a button. It also allows you to auto-deploy code changes to diverse environments (development, staging, testing, QA, production, and so on…) so that companies can quickly troubleshoot and fix bugs and respond to changing business needs.

This challenge of service discovery and CI and CD for a microservices application involves being able to identify where dynamically deployed microservices are deployed, especially in the case of additional instances.

Microservices are composed of a set of separate components and services, each managed by a separate team with an independent lifecycle and an independent CI/CD pipeline.

There are many challenges in the implementation of microservices:

- Low visibility into the quality of changes introduced in each service's pipeline in the application
- Uncertainty about whether each launched pipeline meets security and compliance requirements
- The independence of each pipeline can pose a pipeline control problem – for example, security vulnerabilities, performance issues, a flawed automated testing system, version control, and technological limitations
- Infrastructure duplication caused by multiple services and pipelines

Security and observability challenges

Cloud-native applications present additional challenges for security and risk management because they are inherently complex.

Several independent services to secure

Especially if we're using a combination of containers, Kubernetes, and serverless functions to take advantage of microservices, we'll have multiple services to protect in multiple environments throughout the application lifecycle.

Regular changes in environments

In the agile methodology, teams unveil a new version every week (or even daily, in order to correct a bug, for example). This presents a challenge in terms of the security of what is deployed, which makes the task of security personnel more difficult in terms of taking control of these deployments without slowing down the speed of release each time.

Zero trust and service identity

Unlike monolithic applications that use a physical machine or a virtual machine as a reference point or the stable node of a network, cloud-native applications and, especially, services are deployed in different places. They can even be replicated in several places, providing us with the ability to stop and then restart them at any time. The security of these services requires a network security model that takes into consideration the context of the application, the identity of the microservices, and their networking requirements. This leads us to build a model of zero trust around these requirements.

Zero trust is a strategic approach that consists of protecting organizations by eliminating implicit trust and continuously validating all phases of digital interactions. Zero-trust security is an IT security model that requires strict identity verification for all persons and devices attempting to access resources on a private network, whether inside or outside the network perimeter.

Summary

In this chapter, we learned about application design and lifecycle management, from planning to maintenance. We covered the Twelve-Factor App design methodology, an influential software application design model to build scalable application architectures. Next, we examined serverless applications, also known as FaaS, and finally, we learned about the challenges faced during the implementation of a cloud-native application.

In the next chapter, we will learn about the cloud computing journey, the cloud adoption methodologies, and the best practices, different tools, and resources that can simplify and accelerate your migration to the cloud.

Further reading

For more information about monolithic applications, you can use the following e-book: `https://learn.microsoft.com/en-us/dotnet/architecture/containerized-lifecycle/design-develop-containerized-apps/monolithic-applications`.

If you need to learn more about the microservices approach, you can follow these links:

- `https://learn.microsoft.com/en-us/azure/architecture/guide/architecture-styles/microservices`

- `https://learn.microsoft.com/en-us/dotnet/architecture/microservices/`
- `https://dotnet.microsoft.com/en-us/learn/aspnet/microservices-architecture`
- `https://learn.microsoft.com/en-us/dotnet/architecture/microservices/multi-container-microservice-net-applications/microservice-application-design`

2

Beginning Your Application Migration

In this chapter, we will support you on your cloud computing journey by sharing cloud adoption methodologies and the best practices, tools, and resources that can simplify and accelerate your migration to the cloud. We will introduce the prerequisites to consider before migrating an application to Azure.

In order to apply the best practices to digitally transform and accelerate the operational results of internal applications, at the beginning of this chapter, we will explain the organizational features identified by Microsoft Azure. This will improve readiness for the cloud, and as these features are based on the **Cloud Adoption Framework**, we will plan the migration and the different strategies for cloud migration according to the Azure hosting options.

In this chapter, we're going to cover the following topics:

- Understanding the Cloud Adoption Framework
- Rationalization
- Understanding Azure hosting options
- Evaluating migration considerations

Understanding the Cloud Adoption Framework

Before you start to plan a cloud migration, you need to define a cloud strategy by first creating a business case and then drafting technical plans that will be improved during the first step of the migration. You must make sure that you can run your workload as before, preparing a qualitative model to estimate the migration costs and performing the migration workflow, using a proof of concept as an experimental phase to limit any impact on your organization. You can follow this link to learn more about the Cloud Adoption Framework and seek more guidance on each migration phase in the cloud adoption journey and assessments: `https://learn.microsoft.com/en-us/azure/cloud-adoption-framework/`.

The Cloud Adoption Framework is guidance on the best practices, providing tools, guidance, and narratives to help organizations adopt the cloud and achieve business outcomes. This framework helps organizations define a robust cloud strategy, plan for successful workload migrations, and ensure complete control over their cloud environment. The tools help organizations to develop technology, business, and people strategies to achieve the best possible business outcomes from your cloud adoption efforts.

The Microsoft Cloud Adoption Framework for Azure includes several phases:

1. **Define strategy**: We define a business justification and business outcomes, including a financial considerations evaluation. In this phase, we understand the technical considerations.

2. **Plan**: We align actionable deployment plans with business outcomes. In this phase, we develop a cloud adoption plan to migrate an organization's workloads and a skills readiness plan. You can download the strategy and plan template at this link: `https://raw.githubusercontent.com/microsoft/CloudAdoptionFramework/master/plan/cloud-adoption-framework-strategy-and-plan-template.docx`.

3. **Ready**: We will prepare the cloud environment based on the planned changes. In this phase, we set up the first landing zone.

4. **Adopt**: We will migrate and modernize existing workloads or innovate to build new cloud-native or hybrid applications.

5. **Govern**: Cloud governance is an iterative process. To adopt cloud migration, we need to govern the environment and workloads.

6. **Manage**: We will define the business commitments, operations baseline, and operation maturity.

Now that we have discussed the Cloud Adoption Framework, we will compare cloud migration with cloud adoption.

Cloud migration versus cloud adoption

We need to understand the difference between cloud migration and cloud adoption.

Before building any migration strategy and talking about cloud migration, it is important to understand that any organization will have to change the tasks of its employees – if we are talking about the cloud and its benefits, we need to think about automation, agility, DevOps, different internal processes, security related to data confidentiality, access, legacy, and governance.

Cloud migration is moving workloads that currently run on your local network to the cloud. Cloud adoption means taking advantage of the capabilities of workloads that run in the cloud that are not available when those workloads are running on your on-premises network.

The basic steps to consider before starting a migration

Before starting any migration, we need to follow these steps.

Organizing a cloud team by assigning roles

A cloud team comprises enterprise architecture and innovation leaders. It's like a committee working on a major transformation, from the analysis and definition to the establishment and implementation of all strategies.

Companies that deploy a cloud-first strategy typically opt for **Software as a Service (SaaS)** before considering **Platform as a Service (PaaS)**. In some use cases, **Infrastructure as a Service (IaaS)** – *lift and shift* – is the recommended solution to move on-premises applications to the cloud without redesign. However, depending on your environment and application, you may decide to stick with your on-premises infrastructure, perhaps feeling that keeping everything on-site gives you more control.

Many companies have employees with expertise and skills developed over the years related to a particular custom legacy system, platform, technology, or configuration. Even if this expertise evolves among employees over time, it can become obsolete as the functional area of activity changes. However, these skills are needed so that you can achieve the same results after migration as you had before. Many employees in an organization can contribute – they may be network engineers, system or database administrators or operators, **quality assurance (QA)** engineers, Scrum masters, product owners, security specialists, functional and business analysts, or architects.

If you want to use an agile methodology such as Scrum within your organization, you must start by assigning specific roles to all members selected as potential candidates to work on the cloud migration, or the members who own the products, according to the tasks they will be responsible for in your cloud strategy. Agile development cycles allow you to break your cloud migration execution into a series of small, well-defined, testable, and repeatable processes. This means that every aspect of your cloud migration is done in short sprints so that you can quickly identify and fix issues, optimizing migration speed and efficiency.

You can follow this link to learn more about preparing for technical complexity with agile change management: `https://learn.microsoft.com/en-us/azure/cloud-adoption-framework/migrate/migration-considerations/prerequisites/technical-complexity`.

In conclusion, you need to establish several teams, including a strategic team that works on the strategic planning of the cloud adoption, and the result will be a solid cloud strategy document. Whatever approach is established, this team defines the motivations behind it and the desired business results. It must align the business priorities and the efforts to be made during the transition to the cloud, analyzing any impacts on the business functions or adoption issues. This team can include decision makers such as business leaders across the organization or directors. Basically, this team ensures that cloud adoption efforts progress in line with business outcomes.

Once the strategy has been established and a document has been prepared, including planning, another team should be set up to guarantee the success of migration and adoption, which is the cloud adoption team. This team will define the applications to be migrated without being modified, those that need to be modernized, and those that require new architecture or a complete redesign. These decisions are related to the application and the ability to change it if it needs to be redesigned, for example. These decisions kick off the adoption journey. By the end of their process, the team will have developed skills for different scenarios and good practices.

We cannot guarantee the success of adoption if it is not governed because then we may encounter unexpected problems. We must align cloud adoption with cloud governance functions or the creation of a new team that deals with governance. The cloud governance team collaborates with the other teams and works on the governance baseline assessment, based on conversations with various stakeholders, and deploys the basic governance tools and organizational configurations needed to manage the environment during further adoption-related work.

All of these teams work together throughout the cloud adoption journey.

Understanding your workloads

Once an IT organization has been transformed to build a new structure dedicated to cloud strategies and analytics, we can follow these steps:

7. Define the different roles before creating the core team, highlighting goals, key indicators, and metrics; this step is important, as it considers an organization's workload.

8. Every member of the cloud team needs to work with the IT team to define the infrastructure characteristics – for example, the number of servers, the number of **virtual machines** (**VMs**), storage, the operating system, capacity, the usage rate per hour or per day, the specific configuration to consider, the licenses installed, and any other technologies. Every detail is important when choosing the best cloud model from IaaS, PaaS, and SaaS.

For each organization, the steps can be changed accordingly.

Application and database coupling

Every application can be replicated across multiple servers to ensure continuous availability, and in this step, we have to consolidate all elements to define the links between all the systems, applications, and databases so that we are able to select potential candidates and prioritize all applications in a list.

Following a directive to select and move some of the applications to the cloud

Once priorities have been set, more work is required by all members in terms of data compliance, privacy, confidentiality, and the technology used to decide whether migration can occur without changing, offloading, or redesigning anything.

Building a draft vision architecture for potential candidates

We can select some candidates to be a proof of concept to learn more about the cloud and increase competence. Every application should be tested in the context of a specific use case – it is recommended to select an easy case to be able to best understand the different issues and risks. Therefore, different solutions can be used as a reference for other applications.

Starting trials using the cloud provider's free plans

Once your strategy has been determined, you can select one or more providers to start the proof of concept – you can see the differences between them in terms of costs and services. Each vendor offers a testing service, such as DevTest Labs in Microsoft Azure, where you can use cloud services and test for feasibility and predict problems.

The Microsoft Cloud Adoption Framework for Azure is considered a complete cloud adoption life cycle guide; it is used to identify and put transformation opportunities in order, evaluate the different perspectives of evolution, and above all, improve your preparation for transition to the cloud by following a transformation roadmap that evolves iteratively and gradually. Before talking about this framework, we need to understand the business needs of a company and the different migration scenarios. The first step to migrating an application is really important to build your strategy and successfully migrate according to plan.

Preparation of a plan to change roles within the organization during migration

The preparation of a human resource strategy is really important. As you complete work on your migration, you may need to include a new organization with specific roles that bring old skills together across jobs – you have to consider a development team converted into new roles related to cloud positions, such as an architect or integrator. We move from manual tasks toward a more complete adoption of automation, scaling and provisioning more resources more efficiently.

When following the previous steps, any organization can encounter obstacles during its cloud migration journey, which we can easily avoid by applying common approaches shared by multiple customers.

Cloud Adoption Strategy Evaluator

You can use the Cloud Adoption Strategy Evaluator by accessing this link: `https://docs.microsoft.com/en-us/assessments/?mode=pre-assessment&session=local`. It will help you build a business case to ensure an efficient, well-designed, successful, and easy transition to the cloud.

Figure 2.1 – Cloud adoption strategy assessment

In conclusion, to accelerate migration and for it to succeed, we need to use the methodologies of Microsoft Cloud Adoption Framework for Azure as follows:

1. Define a strategy where the input is illustrated in a document, defining the business outcomes that you want to achieve and your motivations. This document will be used by all contributors during the app migration.

2. Align the cloud adoption plan with the business goals, be engaged, and select a service provider that offers support throughout the process and beyond.

3. Be ready for a cloud environment where many concepts will be added and many processes and ways of working will be changed. Don't forget to collect useful information about your dependencies using automated cloud migration tools. Prepare an inventory of your infrastructure by assessing your on-premises environments. The vendor can provide sizing advice and other necessary services, such as workload cost estimates and performance metrics.

4. Adopt the cloud by migrating everything using *lift and shift*, with or without changes, or innovate to implement the desired changes across IT and your business.

5. Don't forget governance, which is a key factor in a successful migration, since it guarantees data security.

We have assessed a cloud adoption strategy, and now we will discuss cloud rationalization.

Rationalization

Since it's all about the application, your choice of strategy will depend on the migration process. It's really important to understand the different strategies.

Cloud rationalization is a process used to evaluate assets in order to define the best way to migrate or modernize each asset in the cloud.

For more information about this process, you can follow this link: https://learn.microsoft.com/en-us/azure/cloud-adoption-framework/digital-estate/.

In this section, we will talk about the different cloud migration strategies and what are commonly known as the five Rs of rationalization, which can be implemented by organizations to migrate applications to the cloud.

In the planning phase, organizations prepare an appropriate plan and approach for migrating each internal application by prioritizing dependencies. This strategy defines success or failure, as it will maximize the value of the cloud infrastructure while minimizing time, costs, migration risks, and effort.

Simplifying the app is the first step in your migration. Microsoft defines it as "*a process used to evaluate the different assets in order to determine the best way to migrate or modernize each asset in the cloud,*" and the best method to analyze your application environment and choose the convenient path for your migration is using the five Rs of application modernization.

The technical experts in a company must prepare a release plan for each application and need to decide whether they will work across all applications, even if they cannot be integrated into the cloud, or choose a particular application for which they must initiate a strategy.

Let's explore the five Rs strategy for migrating applications to the cloud, presented in the following figure:

Figure 2.2 – The five Rs strategy

Rehost

Commonly known as *lift and shift* re-hosting is the process of moving your applications from an on-premises environment to a cloud environment without modification. It is a simple strategy with minimal risk and effort to adopt because it involves moving from one environment to another, similar to moving from one server to another. This approach is suitable for legacy app migration if you're looking to move quickly and reduce infrastructure and operating costs, if your on-premises infrastructure is costing your business too much, or if you have an IT team with limited cloud knowledge.

This encourages the organization, in the long run, to move to a cloud-first approach to business, and over time, the team's IT skills will grow within the business.

Re-platform

We can move an application to the cloud after applying a few optimizations, and we can make some updates to optimize it and make it better prepared to be deployed to the cloud. With the re-platforming strategy, we can switch applications from a self-hosted infrastructure to managed services. You can minimize the amount of time you spend managing database instances by migrating to a database-as-a-service platform and leveraging Microsoft tools, such as Azure SQL Managed Instance, or migrating your application to a fully managed platform, such as Azure App Service. Re-platforming is also known as the lift-tinker-and-shift method, which was introduced by AWS.

You can learn more about this by following this link:

```
https://docs.aws.amazon.com/prescriptive-guidance/latest/large-migration-guide/migration-strategies.html
```

Refactor or re-architect

If you're planning more application changes, these can be based on cloud templates and by thinking about more solutions, thereby refactoring the code. By opening your **Independent Software Vendor** (**ISV**) up to new business opportunities, you can completely overhaul an application to adapt it to the cloud, and the ISV will have greater cloud efficiency with better access to resources, speed, and optimized costs. If the application is not compatible with the cloud, it has to be rewritten to fit the cloud better. The goal is to transform non-cloud applications, which usually means turning a legacy system into a cloud-native application. Organizations look to modify their monolithic application into a service-oriented architecture, such as by using microservices and going serverless. This strategy boosts agility, scales apps, and improves business continuity, allowing you to adopt a new cloud capability easier by using new technology stacks.

Rebuild

The majority of organizations are moving from on-premises to hybrid models. Some are still developing internally, making themselves compatible with a cloud move at a later stage. However, if we need to remove old apps and redevelop them using cloud features and services, this will require a good command of functional and technical concepts, as well as knowledge of an existing application, business processes, and cloud services.

Sometimes, we want to abandon applications because they do not meet the current needs of an organization, as they are not aligned with the current business process or are missing several elements, so there is no need to invest more. The solution is that the ISV leaves the old application, and it will be rebuilt with a new code base that is aligned with the organization's objectives and a cloud-native approach. Rebuilding involves using Azure PaaS capabilities, such as Azure Functions, Logic Apps, and Azure SQL Database.

Replace or repurchase

The replace strategy means that we will use a SaaS solution instead of the current application.

When building your app from scratch, you have to use the best technology and approaches available at the time, but technologies change quickly, and others can quickly become outdated, can be difficult to maintain, and don't follow industry best practices. It is better to replace legacy applications using a SaaS solution if you don't want to invest in a development team to relaunch your app.

When migrating workloads to the cloud, you can identify redundant or old applications that have been shut down for a long time and can be retired to save costs. The savings will ultimately improve the business case.

In conclusion, we have discussed the different methods you can use to migrate to the cloud:

- Re-hosting – lifting and shifting
- Re-platforming – repackaging your application without major code changes
- Re-architecting – modernizing your code and breaking monoliths into small services
- Rebuilding – completely rebuilding your app in a native cloud
- Replacing or repurchasing – replacing your application with a SaaS solution

We have discussed the different methods to migrate an application and cloud rationalization. Now, we will look at the different options for Azure hosting in more detail.

Understanding Azure hosting options

In this section, we will present the Azure hosting options that we need to consider before migrating to the cloud.

The *lift and shift* model is based on the principle of moving your existing applications or services to Azure VMs, with operating systems and network configurations similar to their current on-premises configuration when deploying in a cloud platform. A successful lift and shift migration leverages the benefits of cloud infrastructure and manageability, while minimizing both migration costs and the time required to complete the migration.

The following figure presents the different options to consider before migrating an application:

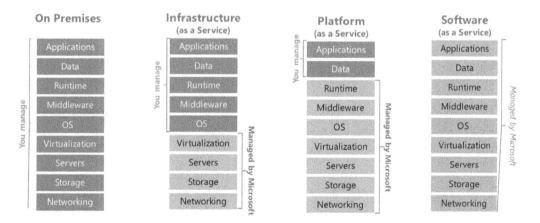

Figure 2.3 – Separation of responsibilities in the Azure model

When migrating .NET applications, or even new applications, to Microsoft Azure, consideration must be given to the choice of compute, database, networking and security, governance, and authentication and authorization.

When migrating .NET Framework applications to Azure, we have several choices; in older .NET versions, we depend on Windows, and we will have these choices:

- **Azure VMs**: These are used for an easy application migration path. If the application has a strong dependency on local `.msi` (or `.exe` or legacy app) installations and the server, we can switch to Azure VMs, where the deployment environment is a VM, so it is similar to on-premises servers.

- **Azure App Service**: If an application has no dependencies on the server and we have a clean ASP.NET web app, such as Web Form and MVC, or an N-tier app, such as a web service, WCF, or a web API accessing a database server, using Azure App Service is a big asset for PaaS maintenance, offering a simple way to manage and scale apps in Azure.

- **Windows containers**: If the application has dependencies on the origin server, these dependencies can be included in the Docker image, and we can modernize the application to be ready for cloud DevOps.

Now that we have discussed Azure hosting options, we will discuss how to evaluate migration considerations in the next section.

Evaluating migration considerations

Understand your application environment so that you can select an application that will have the most impact and benefit and be an ideal candidate for the migration of other applications. Assess your entire application portfolio and migrate only the selected candidate first.

Data security and confidentiality are essential elements; it is very important to ensure compliance with internal policies and legal or regulatory requirements. In the analysis and planning phase, a key step is to know the different types of data used in the service to be migrated, in order to establish a data encryption model and the isolation of user accounts from each other by ensuring well-controlled cross-account access. Risk management leads to decision-making on appropriate security measures for each type of data and also an assurance of data compliance.

In order to avoid surprises and overcome all the challenges when migrating applications and their data, take your time when choosing your migration approach because migration is not a one-time process. Prioritizing your applications or tasks to migrate in order of priority and complexity, working on a potential candidate, and then successfully migrating it can help teams overcome any issues they may encounter afterward and help estimate complexity, time, and cost.

Summary

In this chapter, we first discussed the Cloud Adoption Framework. Afterward, we presented the five Rs strategy for cloud migration: rehost, re-platform, refactor, rebuild, and replace. Then, we explained the different Azure hosting options – IaaS, PaaS, and SaaS – and talked about the migration considerations.

In the next chapter, we will focus on the modernization of existing .NET applications, which means a strategy of moving a workload to a newer or more modern environment without modification, or with minimal modification to the application's code and basic architecture.

Further reading

If you need more information related to application cloud migration strategies, you can follow this link: `https://www.gartner.com/en/documents/1485116`.

If you want to choose the right Azure hosting option, you can learn more by following this link: `https://learn.microsoft.com/en-us/azure/developer/intro/azure-developer-key-concepts?source=recommendations`.

Questions

1. What is the Microsoft Cloud Adoption Framework for Azure?
2. What are the different strategies that we can use for cloud migration?

3

Migrating Your Existing Applications to a Modern Environment

When moving a workload to a modern environment, it is important to accelerate a digital transformation and ensure the safe transition of existing applications. To demonstrate this, in this chapter, we will use the case of a healthcare company that implements and deploys solutions that improve healthcare access to patients as an example, offering diverse applications (mobile, web, and desktop applications) for doctors and hospitals.

The company regularly works through its solutions to improve the experience of patients and healthcare professionals. Patients are always looking for more personalized experiences. The decision to start a digital transformation and migrate applications to the cloud was not so easy, as the impact of this evolution needed to be studied by establishing business continuity planning.

In this chapter, we will focus primarily on the initial modernization of existing Microsoft .NET Framework web- or service-oriented applications, which means the strategy of moving a workload to a newer or more modern environment without modification or with minimal modification of an application's code and basic architecture. We will highlight the benefits of moving applications to the cloud by performing a partial technology modernization of applications, such as the use of Windows containers and the associated compute platforms in Azure that support them.

In this chapter, we're going to cover the following main topics:

- Exploring a migration approach and modernization, scenarios, and paths for an existing .NET application

- Migrating an ASP.NET web solution to an Azure VM

- Migrating an ASP.NET web solution to Azure App Service

- Migrating an ASP.NET web solution to a Windows container

- Migrating a database to Azure

Exploring a migration approach, modernization, scenarios, and paths for an existing .NET application

The decision to modernize a company's solutions does not imply a complete restructuring of its applications. Even re-architecting an application, such as using a microservices-based approach, is not always the right solution due to cost, time, and application constraints. Don't forget to take into account the needs of the company and the requirements of the applications to be migrated in order to be able to optimize the profitability of the established strategy when migrating to the cloud.

In a digital transformation, it is important to understand two points:

- The destination

- Any waypoints during the journey

Note that there are many potential destinations for any application, and cloud computing deployments can be a mixture of the two.

For a traditional .NET application, we have several strategies to adopt to successfully move to the cloud. We need to know whether we want to only migrate, which means taking a gradual or incremental approach when investing in moving assets to the cloud based on business needs, or modernize, which allows us to exploit new technologies, such as containers, microservices, and serverless architecture.

If we consider the business needs of an enterprise and the requirements of each existing application, the most important step is the prioritization of applications. These applications require the following:

- Transformation or re-architecture that delivers modern customer experiences from the existing applications in order to increase their business value and technical responsiveness

- Partial modernization

- Preparation for **lift-and-shift** directly to the cloud

The following figure describes the different paths for moving to the cloud and the preparation of applications ready for a cloud infrastructure.

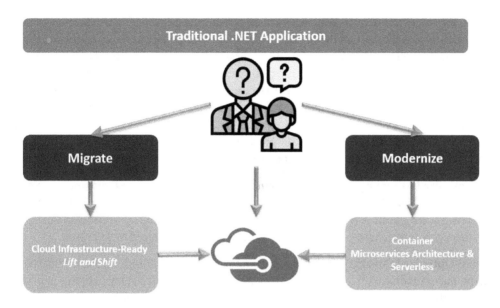

Figure 3.1 – Paths to move existing .NET applications

This figure presents the most common scenarios that occur in the case of a public cloud or even hybrid architecture. These maturity models can be applied not only for legacy monolithic applications but for any architecture style – for example, an **n-tiers** application.

For each maturity level of the migration or modernization process, we determine the following key approaches:

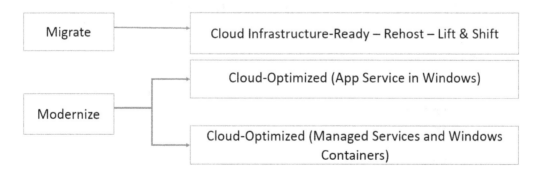

Figure 3.2 – Migrate or modernize?

For **Cloud Infrastructure-Ready**, **Rehost**, or **Lift and Shift**, which is an **Infrastructure-as-a-Service** (**IaaS**) model, if we will rehost all the infrastructure from on-premises to Azure **Virtual Machines** (**VMs**), we can use tools such as **Azure Migrate** (`https://azure.microsoft.com/en-ca/ services/azure-migrate/#product-overview`) to assist the migration, and we need to use other tools such as **Azure Site Recovery** (`https://azure.microsoft.com/en-ca/ services/site-recovery/`) and **Azure Database Migration Service** (`https://azure. microsoft.com/en-ca/services/database-migration/`) to migrate the database. However, we can set up rehosting manually by moving applications to Azure VMs with configuration changes, which is not complicated because it is similar to your current on-premises environment.

Cloud-Optimized (Azure App Service in a Windows Environment) for a **Platform-as-a-Service** (**PaaS**) model is a better solution to use if we want to get the most benefits from migrating to the cloud. No major modification is needed to run the application in Azure App Service, and there is no need to manage infrastructure as we did before because it will be managed by Microsoft.

For **Cloud-Optimized (Managed Services and Windows Containers)** with the PaaS model or IaaS model, no major modification is required, but we need to optimize the deployment in some way to benefit from the advantages of the cloud. We need to add some configuration to deploy on Windows containers, so we need to add support for Windows containers to our existing .NET Framework. Visual Studio includes a Docker support tool that is easy to use. In this case, we can containerize and deploy the Windows container to Azure Web App for Containers or **Azure Container Instances** (**ACI**) with the PaaS Model. Alternatively, we can deploy to Azure VMs with Windows Server 2016 – for example, if we prefer using the IaaS model.

In the next section, we will start with the first approach, the rehost or lift-and-shift. We will migrate an old ASP.NET web solution to Azure VMs.

Migrating an ASP.NET Web solution to an Azure VM

A healthcare company has identified several applications that use different technologies for this strategy. The following applications use versions 4 and 3.5 of the ASP.NET framework:

- The MVC application
- The Web Forms Application
- SignalR
- The web pages
- The web APIs
- **Windows Communication Foundation** (**WCF**)
- SQL, Oracle, PostgreSQL, and MySQL databases

The company has selected a monolithic solution that includes multiple applications. Every application is a legacy monolithic web application.

During this section, we will only migrate two applications. The first application is implemented by using ASP.NET MVC, while the second web application is implemented by using ASP.NET Web API.

This solution is on GitHub: `https://github.com/PacktPublishing/A-Developer-s-Guide-to-Cloud-Apps-Using-Microsoft-Azure/tree/main/Chapter3`.

We will describe the different steps to migrate an old ASP.NET Web solution to an Azure VM. If we need to execute a cloud migration strategy quickly without any significant impact, we will use the cloud infrastructure-ready strategy, a rehost or basic lift-and-shift, that can be automated by using Azure Migrate.

Azure Migrate is a service that provides guidance by analyzing your on-premises VMs. It performs assessments and migrates them to Azure.

In the following figure, we describe a scenario that is based on using VMs in the cloud only for an application and a database server. The existing deployment tools are the same in this case (such as Puppet).

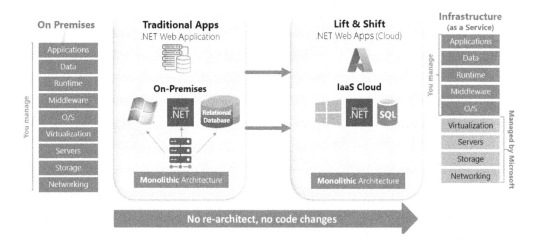

Figure 3.3 – Cloud infrastructure-ready applications

The IaaS model provides an easy transition to the cloud because the application will be deployed to VMs in Azure without any re-architecting, so you have complete control, but this approach is expensive to manage and operate with little cloud value and we need to consider the security risks.

Migrating using Microsoft migration tools and services

In this section, we will present the different tools included in Azure that allow us to migrate an on-premises data center. We can also migrate databases to any Azure service, including Azure Migrate, Azure Site Recovery, or Azure Database Migration Service.

Azure Migrate

Azure Migrate is a core set of services that helps you to migrate your workloads with minimal impact in Azure. It is the best tool to use to achieve your goal and get a return on investment.

In the following figure, we show the different tools in Azure used to migrate an on-premises data center to Azure.

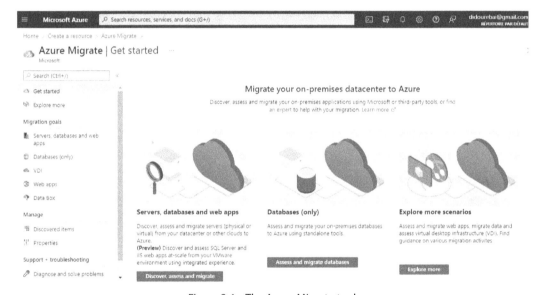

Figure 3.4 – The Azure Migrate tool

With Azure Migrate, we can migrate servers, databases, or web apps. We have also some scenarios to explore before using this tool.

Azure Site Recovery

Azure Site Recovery is also a migration tool used to shift any application or VM to Azure, specifically for hybrid environments. We can use it to replicate an on-premises VM and also physical services to Azure, a secondary on-premises location, or a private cloud data center.

Azure Database Migration Service

It is a tool used to migrate existing on-premises SQL Server, Oracle, and MySQL databases to Azure SQL Database, Azure SQL Managed Instance, or SQL Server on Azure Virtual Machines.

The following figure shows how to select Azure Database Migration Service.

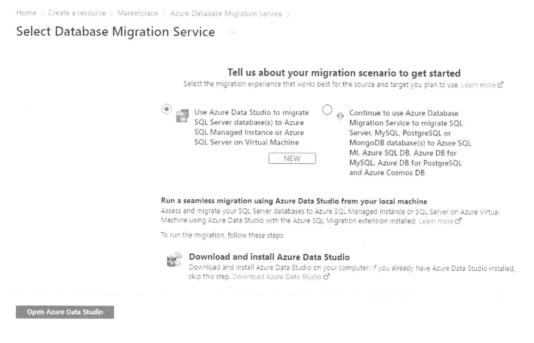

Figure 3.5 – Azure Database Migration Service

We presented the different tools that we can use to migrate our workloads, but we can also migrate our servers manually and deploy applications in Azure VM.

Migrating an application manually

Prior to the .NET Core framework, organizations had the ability to host web applications on their own internal infrastructure in an environment running Windows Server, **Internet Information Services (IIS)**, and SQL Server or another **Relational Database Management System (RDBMS)**. We need to have a similar environment in Azure VM.

To create a VM, go to the Azure portal, and select **Create a resource**. Search for `Windows Server` to create a new VM within a Windows Server environment.

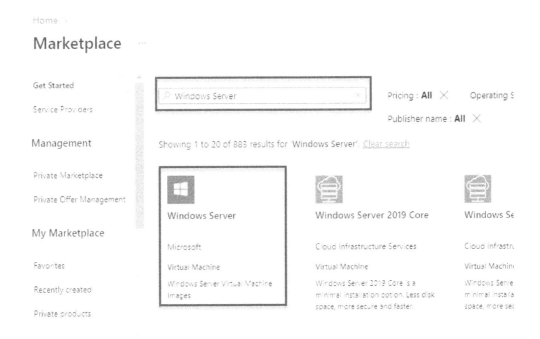

Figure 3.6 – Selecting the Windows Server option

We fill in the information required to create a new VM using an internal image of Windows Server. We select the compute resources size and, as inbound ports, we select a protocol used to access a VM. We can use **HTTP (80)**, **HTTPS (443)**, **SSH (22)**, or **RDP (3389)**, all as presented in the following figure:

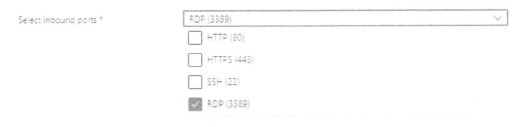

Figure 3.7 – Inbound ports for VM access

We can use **Remote Desktop Protocol (RDP)** to connect and open the VM.

We need to create a new VM in Azure, after which we will lift and shift the application and the data. For an ASP.NET application, we need to enable IIS and provision it in the Azure portal so that we can access it from anywhere. We need to access the hosted sites in IIS from outside of the VM. To summarize, all applications of the company will be moved without any change in technologies and using the same framework; we don't need to upgrade to another recent framework.

The lift-and-shift of on-premises workloads to VMs in the cloud is an easy transition for an organization without spending a lot of time and resources on the migration process, where we move first and we optimize after. Despite the simplicity of this transition, it doesn't require a lot of modification. The ability to move any application without any re-architecting and minimal impact with complete control of workloads make this method an ideal way to begin moving to the cloud. However, it will be expensive to manage and operate, without taking advantage of the cloud for example by setting up a DevOps pipeline. We have to mitigate security risks and a server maintenance process would need to be added. It is better to migrate with a PaaS model if we want to get more benefits from the cloud.

Migrating an ASP.NET Web solution to Azure App Service

In this section, we will work on the applications selected by the company and migrate them to Azure App Service. Azure App Service is a fully managed platform to deploy and manage web apps. We will explore this service in more depth in *Chapter 7, Using Azure App Service to Deploy Your First Application*. This scenario is considered a **cloud-optimized** approach to an application because it is a PaaS solution, and we are able to use more tools for the database – for example, we can use a PaaS solution such as Azure SQL DB, supervise our application by using a monitoring solution such as Application Insights, and ensure continuous integration and continuous delivery using Azure DevOps. This approach helps a company quickly and with little effort migrate to a cloud-native setup, ensuring it will be operable as apps evolve. We will always select Windows as an environment – in this case, we don't need to make any changes.

In the following figure, we show a scenario based on deploying an Azure web app (in Windows).

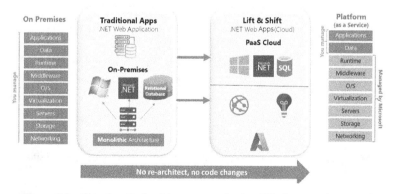

Figure 3.8 – Cloud-optimized (an app service in a Windows environment)

Now that we have discussed migration to Azure App Service, we will see in the next section how we can deploy an ASP.NET application it.

Deploying an ASP.NET application to Azure App Service

The first step before deploying our applications to Azure App Service is to ensure that they don't have any dependencies on features that might not be supported on the platform. We need to complete an assessment using Azure App Service Migration Assistant, which is a local agent.

Migration Assistant is part of the suite of migration tools that helps organizations with their journey to the cloud. Migration Assistant provides a user-friendly interface that is simple to use, like a wizard assistant.

We will follow these steps:

1. Install Azure App Service migration tools (`https://azure.microsoft.com/en-ca/services/app-service/migration-tools/`) on the server.

2. Open the Migration Assistant tool, and it will display the list of sites deployed on IIS. We need to select the application to start the assessment, as shown in the following figure:

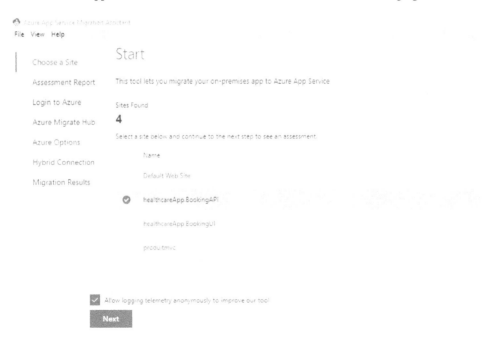

Figure 3.9 – Azure App Service Migration Assistant – step 1 – selecting a site

3. The assessment report will display a list of successful modules, warnings, or errors. We can save the report or pick another site. If we don't have any problems, we can continue to the next step, but if the assistant finds a problem, we need to update the application according to the error message. Sometimes, we need to disable a specific feature and update the application in order to migrate it, and the assistant can guide us during this procedure. If we are not able to update the application and still have many errors, this method is not suitable for this application, and the previous migration to Azure VMs is the required solution:

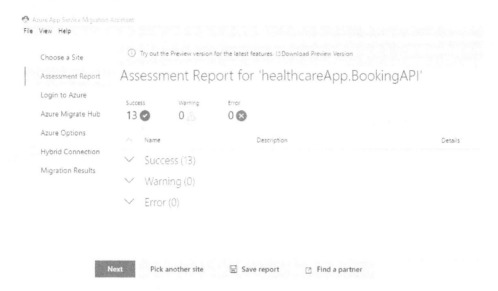

Figure 3.10 – Azure App Service Migration Assistant – step 2 – assessment report

4. We will log in to an Azure account. To do so, follow the steps mentioned in the following screenshot. Migration Assistant will connect to your Azure account:

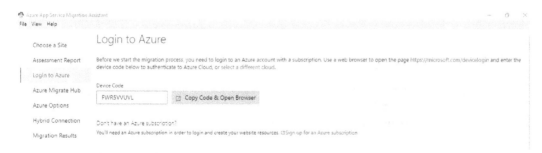

Figure 3.11 – Azure App Service Migration Assistant – step 3 – logging in to Azure

5. If you have created an Azure Migrate project, you can select it, as shown in the following figure:

Figure 3.12 – Azure App Service Migration Assistant – step 4 – Azure Migrate Hub

6. We have to create a new migration project in Azure, as shown here:

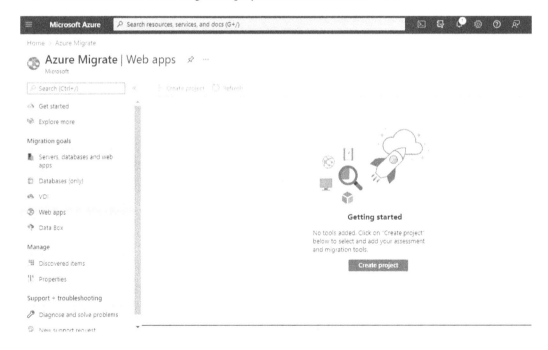

Figure 3.13 – Creating a web app migration project

7. After selecting the migration project, we need to fill in all the required configuration settings to create a new Azure App Service. For the database, in the same window, we will be requested to either set up a hybrid connection to enable database connection or skip setting up a database and use another migration tool to do so:

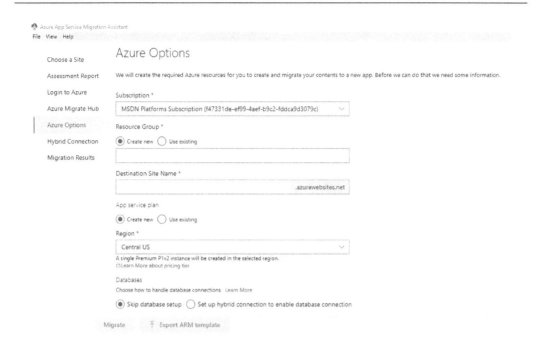

Figure 3.14 – Azure App Service Migration Assistant – step 5 – creating an Azure Web App Service

8. We will select **Set up hybrid connection to enable database connection**, and we will fill in the configuration settings, including the database server and the port.

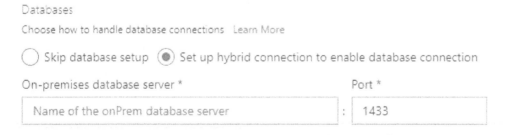

Figure 3.15 – The database settings in Azure Web App Service for hybrid connection

9. Then, the migration process starts, as shown in the following screenshot. It will create the web app and migrate the application. We can export the Azure Resource Manager template, which we will generate using the assistant, and we can use it to automate migration to the web.

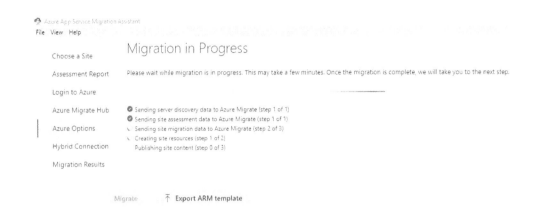

Figure 3.16 – Azure App Service Migration Assistant – step 6 – Azure options

10. Finally, we can check the migration results:

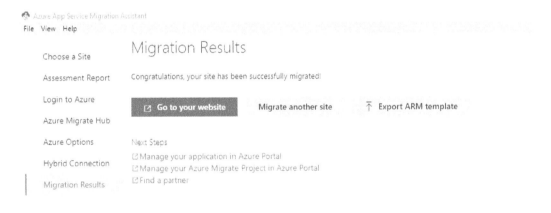

Figure 3.17 – The migration results

We can use the Azure Migrate tool to deploy web APIs and other web applications. You can find out more by following this link: `https://azure.microsoft.com/en-us/products/azure-migrate/`.

Publishing Windows Communication Foundation (WCF) to Azure App Service (Windows)

If we want to deploy any application from Visual Studio to Azure App Service, we can open our application in Visual Studio 2022 – for example, by right-clicking on the project and selecting **Publish**. Select the specific target, which is Azure, and then **Azure App Service (Windows)**. The project will be deployed. We can use this method if we don't have any dependencies or any feature that is incompatible with App Server.

WCF is a framework for building service-oriented applications. WCF allows you to send data as asynchronous messages from one service endpoint to another.

Since the arrival of .NET Core, WCF is no longer supported because it is Windows-specific technology and .NET Core is meant to be cross-platform. However, you can select **Azure** and then **Azure App Service (Windows)**.

For example, in this figure, we will publish our WCF service to Azure App Service.

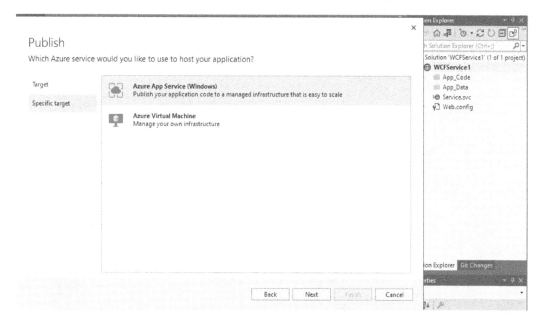

Figure 3.18 – Publishing WCF to Azure App Service (Windows)

We can use a **Google Remote Procedure Call (gRPC)** to host web services inside a .NET core application. To modernize a WCF service, we can use **CoreWCF**; this package allows us to enable existing WCF services and move them to .NET Core.

During this section, we migrated an ASP.NET web solution to Azure App Service and published a WCF service to an App Service Windows environment. In the next section, we will migrate an ASP. NET web solution to a Windows container.

Migrating an ASP.NET web solution to a Windows container

In this section, we take a look at migration, using modern technologies such as containers to deploy ASP.NET Web solutions to containers.

We will learn more about containers in *Chapter 8*, *Building a Containerized App Using Docker and Azure Container Registry*. For now, what we need to understand in our migration with containers is that we will be able to include application dependencies with the application itself, reducing the number of issues that we encounter when we deploy to production environments or test in staging environments. It is the best solution to improve the agility of application delivery.

A containerized application is one of the main pillars of a cloud-optimized application. A container environment offers resiliency, high availability, monitoring using App Insights, and continuous integration and continuous delivery.

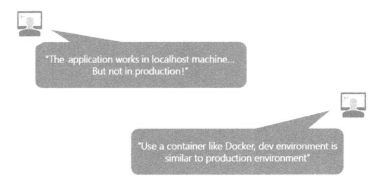

Figure 3.19 – The benefits of using containers

In the following figure, we show a cloud-optimized scenario that is based on deploying on Windows containers (in Windows).

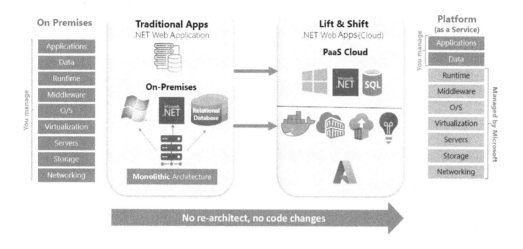

Figure 3.20 – Cloud-optimized (managed services and Windows containers)

A cloud-optimized approach doesn't require you to re-architect your solution or build new code. The total cost of integration is low. It is a new way to accelerate innovation and improve the production environment and your DevOps agility and is a better development operation for a portable app and dependencies. It is also easy to migrate any solution to a container with minimal security risks, but this approach requires more time for your cloud transition and also requires learning skills for building cloud technologies.

Lifting and shifting Windows applications

In this section, we will set up an environment to containerize an application in a Windows environment. We will use Visual Studio 2022, but we can use Visual Studio 2019 or Visual Studio 2017, which both have Docker support.

Setting up the environment

Depending on your application version of .NET Framework, we need to determine the Windows operating system base image. The versions that you can target are presented in this figure. We have only covered the old framework; for .NET Core Framework, we can use Nano Server:

Figure 3.21 – Operating systems to select based on the .NET Framework version

For an old framework starting from 3.5, we can use Visual Studio tools to add support for Docker. The file generated by default is a Dockerfile, which is an easy method to containerize our application without changing code.

Containerizing the app

We can use Visual Studio to add support for Docker. Open your solution, right-click on the application to containerize, and select **Add** and then **Docker Support…**.

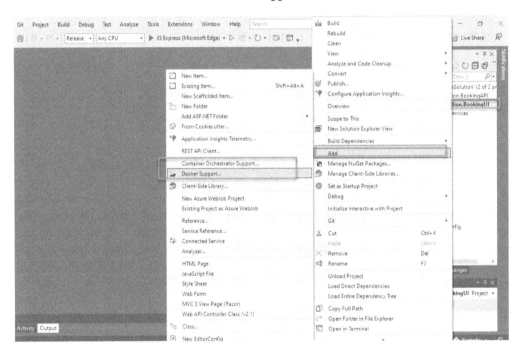

Figure 3.22 – Adding Docker support

A new Dockerfile will be added, and we can see the operating system environment, `mcr.microsoft.com/dotnet/framework/aspnet:4.8-windowsservercore-ltsc2019`, defined in the file as shown in this figure:

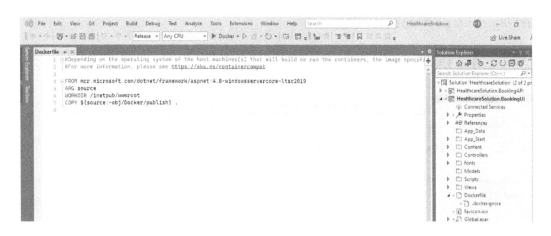

Figure 3.23 – Dockerfile content

If we have more than one container, Visual Studio includes orchestrator support to manage them. To enable this, right-click on the project and select **Add** then **Container Orchestrator Support…**.

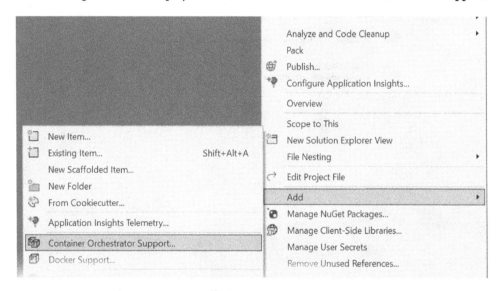

Figure 3.24 – Adding container orchestrator support

Now that we have discussed containerizing an application using Visual Studio 2022 by adding Docker support and container orchestrator support, we will discuss pushing the legacy app in the next section.

Pushing the legacy app

To push an application to any Windows environment, we will open Visual Studio. Then, right-click on the solution and then select one of the following as a location to publish:

- **Azure App Service** as a container (in Windows)
- **Azure Container Registry**
- **Azure Virtual Machine** (with container support)

To deploy our application on an Azure container, we can select an Azure service, as shown in the following figure:

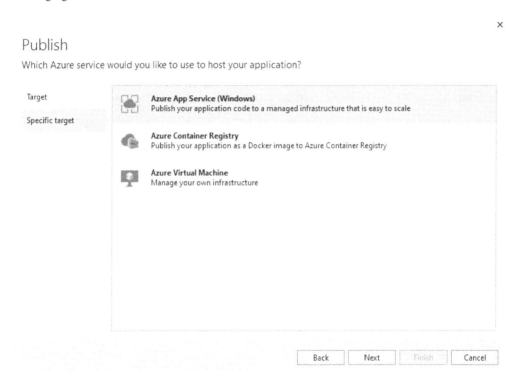

Figure 3.25 – Publishing to a Windows container (IaaS and PaaS)

The next step consists of selecting or creating a new Azure App Service. By selecting the subscription, we can create a new instance or select an existing App Service. In the following figure, we select **Create a new instance** to create an App Service.

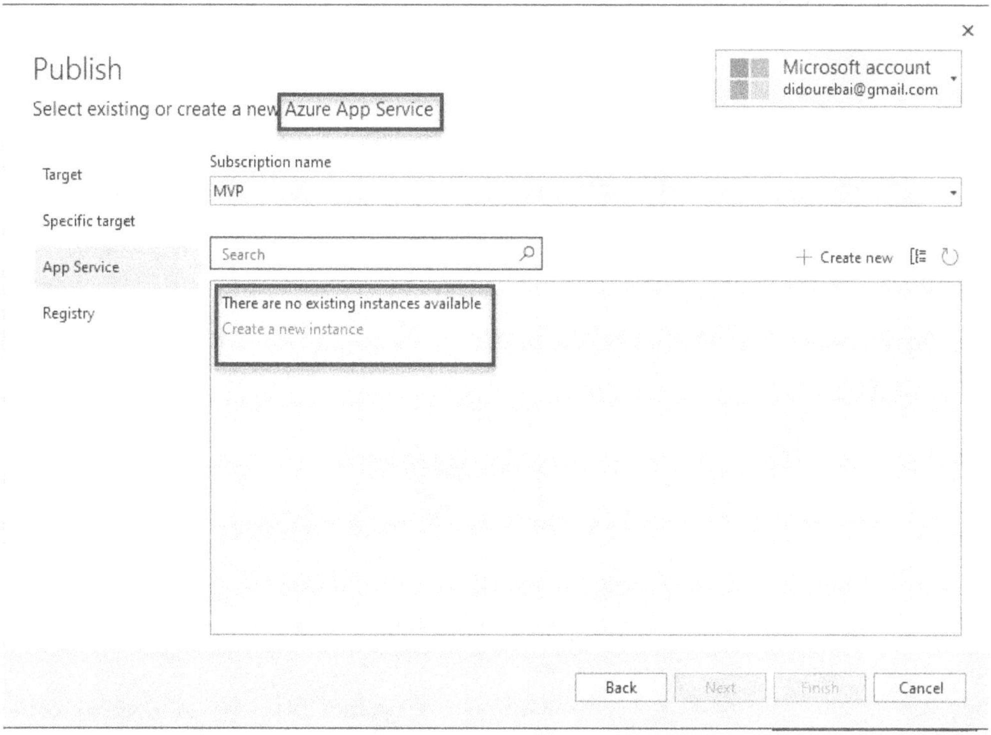

Figure 3.26 – Creating a new Azure App Service window

A new dialog window will be displayed to complete the App Service configuration, including the subscription, resource group, and hosting plan, as shown in the following figure. We will select **Create**.

After completing these steps, you will have an Azure resource and can publish your ASP.NET Core project.

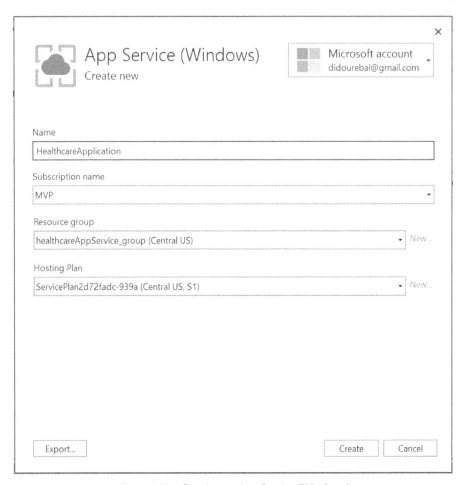

Figure 3.27 – Creating an App Service (Windows)

If we have already created an App Service in Azure, we can select the existing one, as shown in the following figure:

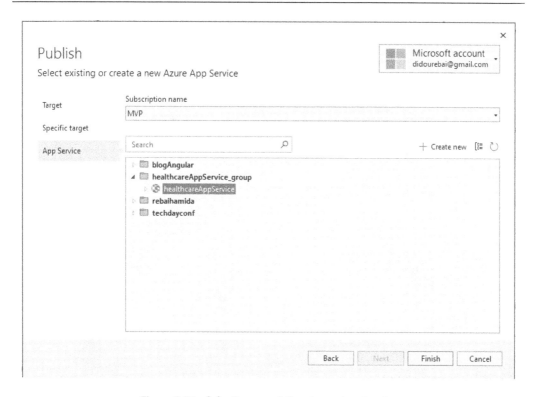

Figure 3.28 – Selecting an existing Azure App Service

We can also select Azure Container Registry or an Azure VM where a containerized environment is configured.

If we want to deploy to Azure Container Registry, we will select **Azure Container Registry** as shown in *Figure 3.29*. In the following figure, we can choose from **Azure Container Registry**, **Docker Hub**, and **Other Docker Container Registry**; we will select **Azure Container Registry**.

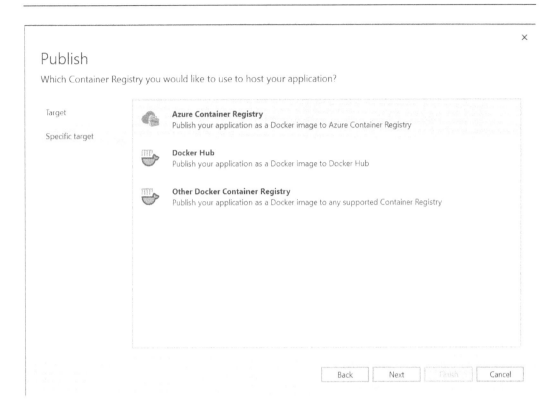

Figure 3.29 – Selecting a container registry

Similar to App Service, we can select an existing registry or create a new instance. We will fill in the required values in the **Azure Container Registry** dialog window.

In this section, we deployed our application in an Azure container, with the application using an old framework and always in a Windows environment. For ASP.NET Core, we were able to deploy in a Linux environment.

Migrating a database to Azure

In Azure, you can migrate your database servers directly to Azure VMs (that is, a lift-and-shift), or you can migrate to Azure SQL Database using the PaaS model. The following figure shows the different relational database migration paths available in Azure and Azure Database Migration Service.

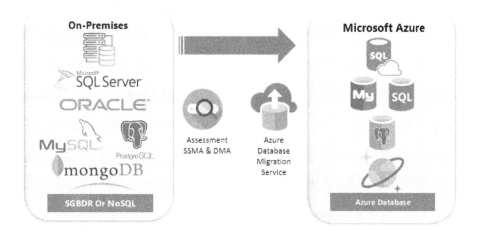

Figure 3.29 – Azure Database Migration Service

In most cases, companies decide to migrate to Azure SQL Managed Instance because it is simple to use with a minimal impact on business, especially if they have an additional requirement for SQL Server instance-level functionality. We can use it as a transition method to move to Azure SQL Database.

Azure SQL Database

Azure SQL Database is a relational **Database as a Service (DaaS)**. It is a fully managed service, making it extremely easy to scale a database, something that was traditionally hard to do with a self-hosted SQL Server. You can also replicate a database in one or more other locations globally, which can improve performance if your application is used worldwide.

Azure SQL Database is available as a PaaS option, as well as a serverless version. The PaaS option requires you to choose a performance level, and you will pay a flat hourly rate to access the database. You can easily scale to a larger or smaller performance plan without disrupting your applications or users by selecting the option in the Azure portal. The serverless option is relatively new and allows you to pay only when you use it; in fact, the database shuts down when you're not using it. This can be ideal for lightly used databases. If you have a developer-only tool that needs a database to run but not 24/7 and that can be available immediately, SQL Database is ideal.

In the next section, we will discuss the migration of SQL Server to Azure SQL Database using Azure **Database Migration Service (DMS)**.

Migrating a database from SQL Server to Azure SQL Database using Azure DMS

To complete this migration, we will use the following:

- SQL Server 2019

- Data Migration Assistant (you can download the application using this link: `https://www.microsoft.com/en-ca/download/details.aspx?id=53595`)

- Azure Data Migration Assistant

- An Azure SQL database – to create a new, select single database in the SQL Database service, you can follow this documentation link: `https://learn.microsoft.com/en-us/azure/azure-sql/database/single-database-create-quickstart?view=azuresql&tabs=azure-portal`

Let's look at the steps:

1. Open Data Migration Assistant and select **New** to add a new project, and then select **Project type** as **Migration**.

2. Select a project name from the dropdown; **Source server type** will be **SQL Server**, and **Target server type** will be **Azure SQL Database**.

3. **Migration scope** will be **Schema and data**. This scope defines whether we need to migrate schema and data, schema only, or data only.

4. Select **Create** to start the migration process:

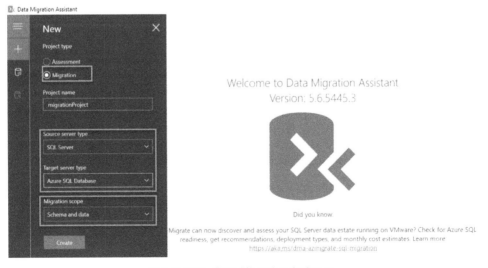

Figure 3.30 – Data Migration Assistant

In the migration process, we will follow these steps:

1. Under **Select source**, connect to the source server using the authentication type and connection properties, as shown in the following figure. Select **Connect** to extract the databases hosted in the source server:

Figure 3.31 – Selecting a source in Data Migration Assistant

2. Under **Select target**, connect to a target server by creating a new Azure SQL database before adding all the settings:

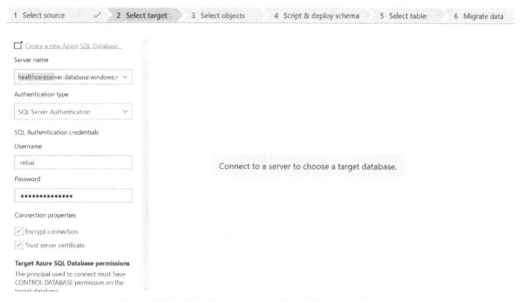

Figure 3.32 – Selecting a target in Data Migration Assistant

3.   Select objects, the script and deploy schema, and the tables, and finally, migrate the data.

Alternatively, we can use Azure DMS instead of Data Migration Assistant.

We will follow these steps to migrate a SQL database to Azure DMS:

1.   Select **Azure Database Migration Service** from the dropdown in the Azure portal:

Figure 3.33 – Azure Database Migration Service

2.   Select the migration scenario and **Database Migration Service**:

Select migration scenario and Database Migration Service

Tell us about your migration scenario: the source database type and target database type, and based on that we will recommend the best Database Migration Service.

Source server type *	SQL Server
Target server type *	Azure SQL Database

Database Migration Service

Database Migration Service *	Database Migration Service

Migrate your databases using Azure Database Migration Service
Azure Database Migration Service is a fully-managed Azure service that helps you easily migrate various databases to their corresponding Azure data services. Learn more

The following source to target migration scenarios are supported:

Source	Target
SQL Server	Azure SQL MI, Azure SQL DB
PostgreSQL	Azure DB for PostgreSQL (Single server or Flexible server)
MySQL	Azure DB for MySQL (Single server or Flexible server)
AWS RDS	Azure SQL MI, Azure SQL DB, Azure DB for MySQL (Single server or Flexible server)
MongoDB	Azure Cosmos DB

Select

Figure 3.34 – Selecting a migration scenario in Azure

3. Create a migration service by introducing the different settings:

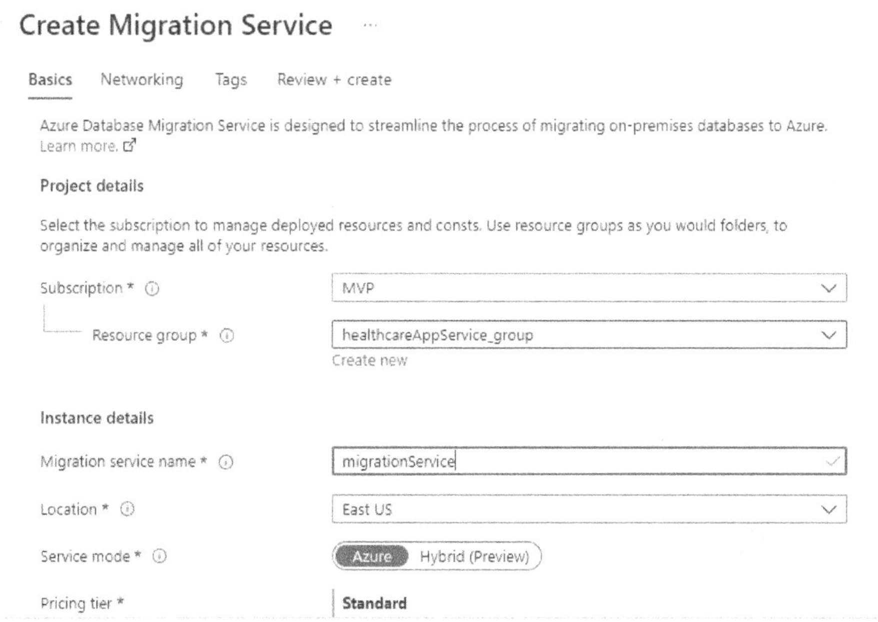

Create Migration Service   ...

Basics   Networking   Tags   Review + create

Azure Database Migration Service is designed to streamline the process of migrating on-premises databases to Azure.
Learn more. ⬈

Project details

Select the subscription to manage deployed resources and consts. Use resource groups as you would folders, to
organize and manage all of your resources.

Subscription * ⓘ	MVP ⌄
Resource group * ⓘ	healthcareAppService_group ⌄
	Create new

Instance details

Migration service name * ⓘ	migrationService
Location * ⓘ	East US ⌄
Service mode * ⓘ	Azure Hybrid (Preview)
Pricing tier *	**Standard**

Figure 3.35 – Creating a migration Service in Azure

4. Complete the migration by selecting the source and the target using the migration wizard:

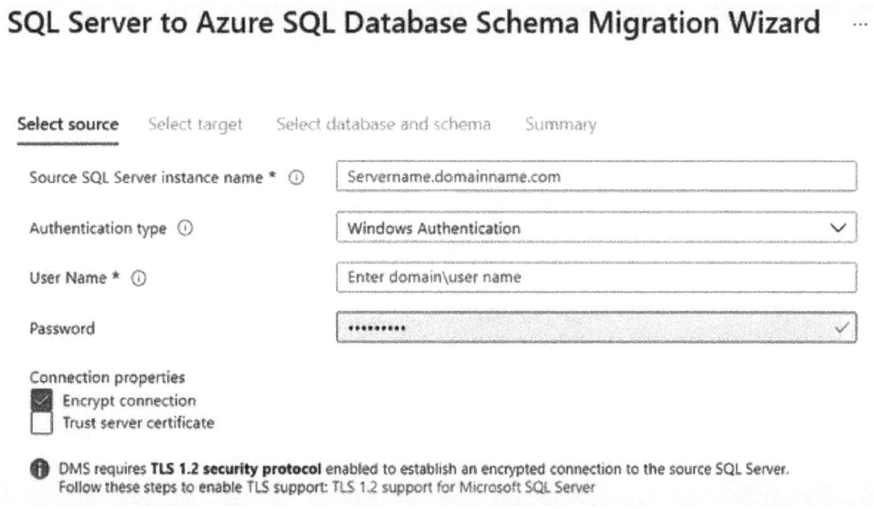

SQL Server to Azure SQL Database Schema Migration Wizard   ...

Select source   Select target   Select database and schema   Summary

Source SQL Server instance name * ⓘ	Servername.domainname.com
Authentication type ⓘ	Windows Authentication ⌄
User Name * ⓘ	Enter domain\user name
Password	●●●●●●●●● ✓

Connection properties
☑ Encrypt connection
☐ Trust server certificate

ⓘ DMS requires **TLS 1.2 security protocol** enabled to establish an encrypted connection to the source SQL Server.
Follow these steps to enable TLS support: TLS 1.2 support for Microsoft SQL Server

Figure 3.36 – SQL Server to Azure SQL Database Schema Migration Wizard

If you need more information about SQL Server migration, you can read this documentation link: `https://learn.microsoft.com/en-us/azure/dms/tutorial-sql-server-to-azure-sql`.

In this section, we discussed the migration of a database to Azure and explored the migration of SQL Server to an Azure SQL server service, using Data Migration Assistant and Azure DMS.

Summary

In this chapter, we explored the migration approach, modernization, scenarios, and paths for an existing .NET app. Then, we learned about migrating an existing ASP.NET Web solution to different scenarios – with an Azure VM, Azure App Service, and a Windows container. Finally, because we can't migrate an application without data, we explored the different options to migrate a database to Azure.

In the next chapter, we will discuss a solution architecture that includes different use cases that we will use in this book. We will discuss the difference between a monolithic approach and microservices and the challenges and solutions for distributed data management.

Further reading

To read more about Data Migration Assistant, you can visit `https://learn.microsoft.com/en-us/sql/dma/dma-overview?view=sql-server-ver16`, and you can follow this link to learn more about Azure Migrate: `https://learn.microsoft.com/en-us/azure/migrate/`.

4

Exploring the Use Cases and Application Architecture

This chapter will take you through a sample .NET Core reference application.

We will use this solution reference in every chapter of this book. It includes a web application, web APIs, and a mobile application. This reference application can be used in different countries, and several different features will be described in this chapter.

We will discuss the solution architecture; we will explain the difference between a monolithic approach and microservices.

In this chapter, we're going to cover the following main topics:

- Healthcare sample solution use cases

- Monolithic versus microservices

- Challenges and solutions for distributed data management

- Solution architecture

Healthcare sample solution use cases

In this section, we will discuss a healthcare company and its different use cases. The company is deploying optimized solutions to improve healthcare access for patients, and it offers multiple applications (mobile, web, and desktop applications) for doctors and hospitals.

The company offers an online service that allows people to find and book face-to-face or telephone appointments for medical or dental care. It offers doctors an interactive platform to manage their appointments, send emails or SMS, manage patient records, and send medical prescriptions to the pharmacy.

Their scheduling system can be accessed by subscribers both as an online service on a website and mobile application and via the deployed office calendar software. The API of the calendar is integrated with many websites.

The sample system is a web application and a mobile application. It also offers an online drugstore and several services, and customers must be authenticated to purchase their products; these products can be shipped by post or picked up in the store.

More solutions are available with the company's bedside patient engagement solutions for hospitals.

Some of the applications are shared to showcase the architecture patterns and to explain more about the use of Azure services. The source code is shared on GitHub: `https://github.com/PacktPublishing/A-Developer-s-Guide-to-Cloud-Apps-Using-Microsoft-Azure`.

The source code in GitHub is not for production use. The technologies used in this solution in GitHub are as follows:

- ASP.NET Framework 4.5

- WCF and ASP.NET Web API 2.0

- NHibernate **object-relational mapping (ORM)** to map with a database

- Xamarin, a native Android and iOS for mobile application

- Partner services to send SMS and notifications

- Basic monitoring with open source tools

The company uses Azure Active Directory for internal users, who are doctors or any practitioners using the internal dashboard, and basic authentication for patients.

Monolithic versus microservices

In this section, we will discuss the difference between monolithic architecture and microservices architecture.

Before talking about microservices and their different benefits, we need to understand the difference between monolithic architecture and microservices architecture.

Monolithic architecture

There are two dominant styles of software architecture among software developers and architects today. In monolith architecture, components are coupled to each other and distributed as a single package. But with this architecture, we have several advantages to consider. Development is really simple because all the code is in one place, so it's easy to find things. The deployment is also simple because when you come to deploy it to production, well, there's just one application to update, and the communication between the components takes place in the same process. It is, therefore, very easy to plan communication between components.

If we only build small solutions, this architecture is recommended because we have several benefits, and these benefits are fairly substantial. But if the application will include more modules and increase in size, it will be hard to maintain it because if a component changes, the other components will be impacted. Likewise, the deployment will be more complicated because, for every change, we have to redeploy the entire system. And when we want to scale the application, it's a simple task to do because the only way to scale this type of application is to replicate the entire application over multiple servers or virtual machines. We can't scale it out horizontally. That's where we add additional servers. However, the only option is to scale vertically, where we provision much more powerful and expensive servers.

For this reason, the hardware requirements of adding servers or virtual machines (on-premises, in the cloud, or using a hybrid approach) will increase, and this will impact the cost of the project. Let's talk about the development tasks in detail. In a large system with a monolithic architecture, it's very complicated to accurately organize development teams, since the design doesn't allow them to work independently. The agile approach will be complicated to apply as a working methodology to organize the different members of the project.

What if we've got more than 20 developers, and they're working for a long period of several years building a system that will have thousands of users and store massive amounts of data? If a code base grows larger, it tends to become more difficult to maintain due to increasing complexity and the accumulation of technical debt. Assuming you've gone to great lengths to keep your monolith modular, you often find that these modules end up becoming very tangled and interdependent. Another issue to consider is when programmers try to deploy their changes in production, even a single-line code change requires the entire application to be redeployed. This means that we require a considerable technical and operational effort, which is risky and usually involves a period of downtime – something that's becoming less acceptable in the modern era of cloud services that are expected to maintain high availability.

Typically, all of the code is in a single code base and all developers collaborate together on that same source code repository. The build artifact is usually a single executable or process that runs on a single host server or virtual machine and persists all of its data into a single database. In a development environment, we use a single technology, such as the programming language or SDK, throughout the code base – for example, C#, Java, .NET Framework, or J2EE.

In some cases, with monolithic architecture, you can very easily find yourself coupled to legacy technology because monolithic applications are often called legacy applications (and vice versa), but it is really important to understand that the two concepts are different. Many legacy applications are monolithic applications, but the term "legacy" actually refers to the development state.

Legacy apps are usually not actively improved but maintained well enough to keep them working for the people who depend on them. Legacy apps end up being retired, either because restricted development imposes functional limitations on users, or because the operations team decides they don't need to be maintained.

Whatever tech stack you built, the original version is going to be very hard to get away from, as you have to upgrade the entire application to move to a newer framework. This reduces your ability to adopt newer patterns and practices or to take advantage of innovation, such as new tools and services that would benefit your application.

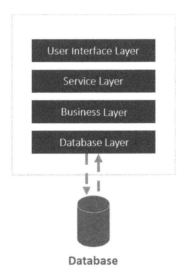

Figure 4.1 – Monolithic architecture

You can think in the previous figure that your application isn't a monolith because it includes several layers. The application is composed of multiple services that can run on different hosts and communicate with each other over the network. However, it's very important to understand that just because you have a service-based architecture doesn't mean you're using microservices.

Microservices architecture

Conversely, when building a new application today, microservices are the preferred software architecture, and a lot has been documented about this already.

You can refer to these links to learn more about microservices: `https://microservices.io/` and (from the Microsoft Learn documentation) `https://learn.microsoft.com/en-us/azure/architecture/guide/architecture-styles/microservices`.

The idea of breaking down a system into small parts is not a new approach. Due to the democratization of the cloud coming alongside innovation in development platforms, microservices architecture is nowadays a good option in certain scenarios. This type of architecture proposes to have lightweight decoupled services with a single responsibility isolated from each other, with independent storage and autonomous deployability.

Microservices architecture has many advantages, and one of the benefits is that it enables improvement in the operation, development, and maintenance of software. Making changes in solutions is no longer a problem for project development. They are a crucial element when we want to incorporate notions of DevOps in an organization. At this point, we will mention some technologies to use with microservices, such as containerization. A microservice should be independently deployable, which means it needs to have a clearly defined and backward-compatible public interface. It provides consistency in deployment, as it allows applications to be packaged as lightweight containers that include all the dependencies required by the service in question. In the same way, it provides process isolation and, therefore, the possibility of adequately scaling your system. For these reasons, the use of containers is often used in conjunction with microservices architecture.

Microservices give us the choice to adopt new technologies without having to upgrade everything all at once, and the flexibility to choose the right tool for the job. For example, one microservice might use a relational database to store its data while another uses a NoSQL database. One microservice might be written in a functional programming language while another might use an object-oriented approach.

The question is, *how we can select the right architecture?*

The answer is not magical and there is no recipe to follow to choose from; both architectures have pros and cons. Some things can be more challenging with a microservices architecture. Knowing and understanding these pros and cons will help you make the best decisions when designing software so that you can decide whether microservices are a good fit for your application, and you can also be prepared to deal with any challenges as they arise. Do not underestimate the complexity of microservices in implementation and testing; if we divide an application into sets of dozens of microservices, the interactions between these microservices can become very complex, making it difficult to understand the behavior of the system as a whole, which can also lead to performance issues.

A monolithic approach can work for small applications. The following figure shows how a microservice should be autonomous, which means it owns its own data:

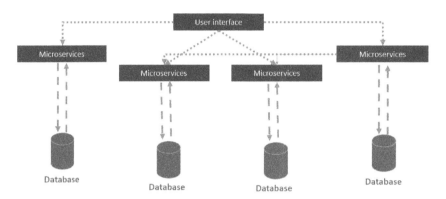

Figure 4.2 – Microservices architecture

Let's explore the different challenges and solutions for distributed data management.

Challenges and solutions for distributed data management

When we talk about large-scale web applications, we have to consider the storage backends to scale and support concurrency.

Through scalability, we aim for incredible data capabilities and very high read/write speeds. The model application server processes a huge number of applications at the same time. Since they rely on background storage systems, they also face very concurrent access at this level.

Performance and capacity expansion can only be achieved sustainably through the horizontal scale mechanism. Even with special hardware equipment, a single database server can only be scaled to a specific load. As a result, we need a parallel distributed system for scalable backend storage.

Background storage needs to maintain the status of an application, so we also expect some consistency of data when reading/writing data into it.

Generally, in distributed systems, failures are often to be expected in advance. If a single node fails, we still want the overall storage system to continue to operate transparently and always be available.

Likewise, we have the ability to perform storage operations within web requests, ensuring low latency of operations.

Let's talk now about the challenges versus the solutions to adopt for distributed applications. Every microservice is autonomous, with all the benefits and challenges that come with it, and every application includes multiple microservices; we have to minimize the coupling between them. We have to identify boundaries for each business microservice and its corresponding domain model and data, in the same way that we can identify boundaries between many application contexts, with a separate domain for each context. We constantly try to keep the coupling between those microservices at a minimum.

Another challenge is related to query implementation, where we will retrieve data for every microservice and hope to avoid unnecessary communication with microservices from remote client applications. To resolve this problem, we can use an API gateway to manage the communication between multiple microservices that have different databases. Another architecture solution is the use of **Command and Query Responsibility Segregation** (**CQRS**) patterns, where we can generate a read-only table with data. This data is used by multiple microservices. This pattern has become almost as well known as the concept of **Domain-Driven Design** (**DDD**). CQRS is about splitting a single model into two – one for reads and the other one for writes.

If you need more details about CQRS, you can follow this link: `https://learn.microsoft.com/en-us/azure/architecture/patterns/cqrs`.

We have more challenges to face, such as achieving consistency across many microservices and planning the designing of the communication between microservice boundaries.

In the next section, we will explore the solution architecture that we will use in this book.

Solution architecture

Whether you're planning to migrate an existing application to microservices or adopt a microservice architecture from scratch, there are some important architectural decisions that you need to make upfront. In this section, we'll discuss some of the key architectural principles for how we define a microservice and its components.

The use case presented previously has a monolithic solution that uses one SQL Server database, with multiple backups every day. The services share the same data model because every web API service targets the same data:

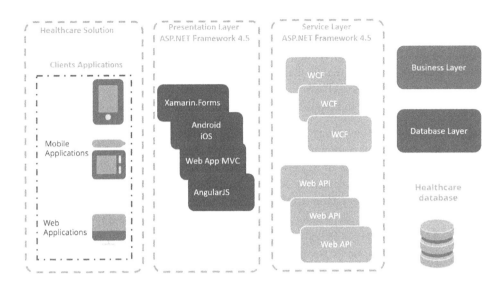

Figure 4.3 – Monolithic solution architecture

Our healthcare solution uses a traditional three-tier application architecture; the client applications include mobile applications and web applications. The presentation layer includes .NET Framework 4.5, and we also have Xamarin.Forms, Web App **Model-View-Controller** (**MVC**), and AngularJS, as well as mobile applications built using Android and iOS. The service layer includes web APIs and web services using **Windows Communication Foundation** (**WCF**). The business layer is the heart of the solution, and it includes the different features of the application. The database layer is used to interact with the database. It is an abstraction layer handling all the data needs; in this layer, we can use ORM internally for any RDBMS communication.

To ensure the availability of the solution to customers in every location, the database is cloned as well as the solution. All applications are still run on-premises, and some of them use classic stacks and application servers.

We need to migrate this application to the cloud, and we need to build a cloud-oriented application. Application modernization is the process of enhancing your current application to deliver a better customer experience and increase **Return on Investment** (**ROI**).

Other reasons to proceed with modernizing applications include having a better ability to achieve faster time to market, because the deployment of new features inside a monolithic application is sometimes quite complex and more time-consuming. The desire to innovate and deliver compelling user experiences continues to drive the need to modernize applications. Another reason is to reduce costs for resource usage, license costs, and support costs.

Summary

In this chapter, we presented a solution use case that will be used during the following chapters. As we will start from a monolithic solution, it was important to understand the difference between this and microservices architecture to be able to select the best approach and re-architect it, using native-cloud technologies. We learned about the challenges and solutions for distributed data management, and finally, we presented the solution architecture.

In the next chapter, you will learn about the different patterns and technologies in the cloud.

Part 2 – Building Cloud-Oriented Applications Using Patterns and Technologies in Azure

In this part of the book, we will focus on technologies and patterns for building cloud-oriented applications.

This part comprises the following chapters:

- *Chapter 5, Learning Cloud Patterns and Technologies*
- *Chapter 6, Setting Up an Environment to Build and Deploy Cloud-Based Applications*
- *Chapter 7, Creating Azure App Service to Deploy Your First Application*

Learning Cloud Patterns and Technologies

This chapter will give you an in-depth exploration of the different cloud design patterns that are useful for building reliable, scalable, and secure applications in the cloud. Before talking about the different design patterns, we will cover the different challenges in cloud development and the different technologies. These patterns and technologies are important to understand.

In this chapter, we're going to cover the following main topics:

- An introduction to patterns and technologies
- Using cloud design patterns
- Using different cloud technologies

Why should you care about design patterns? Why are software design patterns useful? Design patterns describe the objects that interact with classes used to resolve a common design problem in a specific context. These patterns are recipes and templates for best-practice solutions to common problems. By using them, you benefit from all of the experience that went into forming the solutions. This will ultimately result in better-designed and built software in the end.

An introduction to patterns and technologies

Almost every company now has at least some cloud-based services. They could be using a web-based email service, developing a mobile app, or even running their entire company in the cloud. As organizations move to the cloud, they need to decide about protecting information used in their own data centers in a much more distributed environment.

A design pattern is the best practice and approach for a specific issue in software development. It is a set of tools and solutions to help solve common software design problems. It's also an example of enterprise architecture standards. It will also cover a variety of cloud pillars or application domains, such as performance, scalability, and resilience, as this is where the problems are going to occur when you're trying to run an application in the cloud.

In the future, our applications will live in the public cloud. There, they will be exposed to completely different risks and requirements than they are used to when they were on-premises. There is also a strong trend toward building applications with a microservices architecture. This means that we no longer have monoliths but different small objects communicating with each other, sending notifications, alerts, and messages. In addition, we also have a stakeholder expectation that increased significantly over the past few years in terms of how resilient, how fast, how performant, and how cost-effective an application has to be as a whole. This new situation and this new expectation must be reflected accordingly in modern design patterns.

Design patterns aren't algorithms. They are templates, or recipes, that can be used to design your application. The implementation details will be different in each situation.

Your software contains a lot of design patterns already, even if you didn't explicitly use any. You can analyze solutions from different perspectives and think about them in new ways if you define them as design patterns.

The following table summarizes modern patterns and technologies to be used, such as containers. There are many models, most of which are linked or come from domain-driven design.

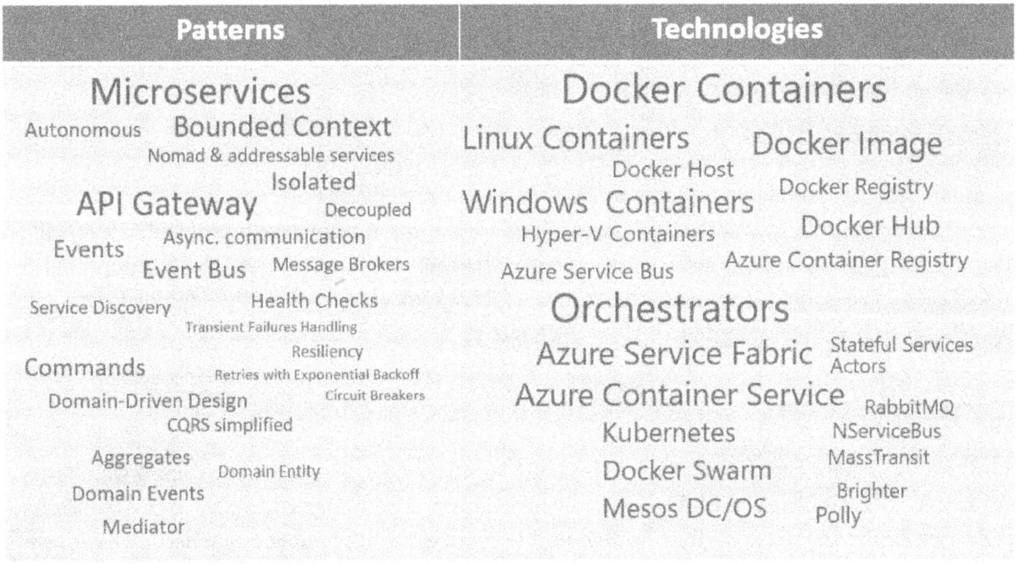

Figure 5.1 – Patterns and technologies in the cloud

In the next section, we will define some cloud design patterns and how they are used according to a specific problem.

Using cloud design patterns

The cloud market has changed dramatically, and cloud application development is on the rise. Since the emergence of the cloud industry, it is no longer just important to reduce IT costs but also to gain advantages in terms of competition, interact directly with customers in real time, and transform a business.

Cloud adoption challenges can come in many shapes, sizes, and severities, depending on the organization.

We will start talking about the different challenges in cloud development.

Interoperability and portability

By definition, interoperability is the ability to write portable code that can work in multiple environments and also cloud providers. Interoperability and portability are the most important features of any cloud environment. In a multicloud context, the different applications and components should be able to run in any cloud environment.

We need to adopt cloud computing standards and best practices to ensure interoperability.

Scalability

Ensuring the scalability of your cloud-based apps is one method to reduce a budget to be more efficient and achieve performance goals. Even though cloud providers can offer scalable services, we need to consistently guarantee our application's availability and ensure that the main server can handle peaks and data load.

We can opt for a hybrid cloud that can scale up and down according to our changing needs to benefit from scalability and flexibility.

Performance issues

In a multicloud or hybrid model, we need to evaluate the latency between the different data centers, as we can have latency when we render data, a user interface, and a style sheet, for example, or any element related to an application. The definition of good performance in an application's user experience is being able to provide a quick response to every user request; otherwise, you will drive your users away.

To overcome this challenge, we need to work with providers and make sure that communication between the application and data does not suffer in the long run. You can start with a **proof of concept**, with monitoring to study the communication between other applications with data, for example. This can help you to determine whether applications can be safely migrated to the cloud. Therefore, it is essential to ensure end-to-end performance testing. There will be no scalability in some situations.

We can have multiple scenarios. In *Figure 5.2*, we will present a common scenario that is always present during a cloud migration. For example, a solution can be deployed in Azure, user interfaces and services can be built using **ASP.NET**, and the solution can use more than one database, such as an **Oracle database** and **SQL Server database**. In the migration analysis phase, the team can decide to keep some sensitive data on-premises for security issues. Some companies use a specific technology such as **Oracle Exadata**, which is a database machine designed by Oracle. It is a database appliance supporting a set of database systems, such as **online transactional processing (OLTP)** and **online analytical processing (OLAP)**. This technology is supported only by Oracle and only for cases where we decide to move to the cloud without any change or database conversion. We will use the cloud service of Oracle; otherwise, the migration will be more complex or expensive. You can see more details at this link: `https://www.oracle.com/ca-en/engineered-systems/exadata/database-service/`.

To interact with a database and the different tools or applications using it, we can have latency. It is really important to do some bandwidth tests on various sites.

In the following figure, we will present the scenario of an application and its data deployed in different data centers:

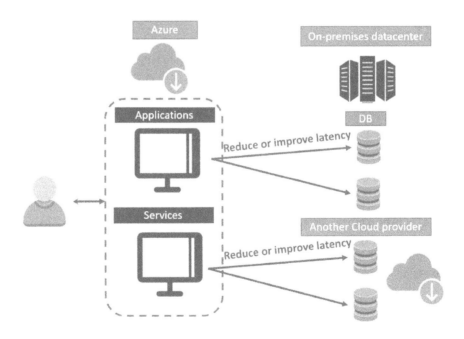

Figure 5.2 – Scenario of the application and data deployed in different data centers

This scenario is not the best solution to avoid latency because we recommend always having data and applications on the same server or provider.

Reliability and availability

One of the advantages of applications hosted in the cloud is the ability to ensure mission-critical operations with minimal downtime, even in the event of disasters.

However, cloud providers can have outages. It is, therefore, important to monitor the services provided using monitoring tools such as Azure Monitor. For this reason, we can supervise our usage, **service-level agreements** (**SLAs**), robustness, and performance of the workloads, including the application and data. When we start any migration, we need to ensure the business dependency of every service and that our cloud-based apps are available 24 hours a day, 7 days a week.

Loosely coupled service design

Customers keep changing their requirements and needs when implementing a product, so it is important that cloud applications are not tightly tied to the underlying service logic and implementation.

By definition, when a component has some knowledge about another component, a **loose coupling** occurs. When components depend on each other, this is the opposite case, called **tight coupling**. Let's consider two classes: the first class changes and the second is affected, so these classes are tightly coupled. We can make changes in these classes to be uncoupled, but they can still work and interact together; if we make changes to the first one, the second will not be affected, and they are, in this case, independent. We can add a new class that will include only things that are loosely coupled, so we can reduce the dependency between the two classes. Loose coupling reduces the dependency between the different components, so we have to architect our solution and consider this, which helps us to build solutions that can handle change well.

Application security

When we move to the cloud, we need to consider security at the beginning, especially when we build an application. We can use authentication and authorization for this. You can use this link for more information (secure code development ISO27001): *ISO 27001 Annex A.14 - System Acquisition, Development and Maintenance (isms.online)* (`https://www.isms.online/iso-27001/annex-a-14-system-acquisition-development-and-maintenance/`). We always need to improve client-side security due to the state of browsers and ensure that critical security should be performed outside the browser. Also, we have to encrypt data using strong encryption. We can use the SSL protocol, for example, or more best practices to meet security needs.

Cloud design pattern

We will now present the different design patterns:

Ambassador Pattern	Create a helper service, which will send multiple network requests on behalf of a consumer service or application.
Anti-Corruption Layer Pattern	Create a class that will be a façade or an adapter layer between the legacy system and your cloud-oriented application.
Asynchronous Request-Reply Pattern	Ensure the decoupling between the backend processing and the frontend host. If the backend requires asynchronous processing, the frontend will require a clear response.
Backends for Frontends Pattern	Create different backend services for different frontend applications or interfaces.
Bulkhead Pattern	Separate application components into pools so that if one fails, the others can keep going. This pattern is used to create applications using a resiliency concept.
Cache-Aside Pattern	Data from a data store can be loaded into a cache on demand.
Choreography Pattern	Instead of relying on a central orchestrator, just let the service decide when and how a business operation is processed.
Claim Check Pattern	To avoid overloading a message bus, divide a large message into a claim check and a payload.
Compensating Transaction Pattern	Reverse the work of several stages that, when combined, produce an eventually consistent action.
Competing Consumers Pattern	Allow several consumers to process messages received on the same messaging channel at the same time.
Deployment Stamps Pattern	Multiple copies of application components, including data stores, should be deployed.
External Configuration Store Pattern	Transfer configuration data from the application deployment package to a central location.
Federated Identity Pattern	Assign authentication to a third-party identity service.

Gatekeeper Pattern	Use a dedicated host instance to protect applications and services by acting as a broker between clients and the application or service, validating and sanitizing requests, and passing requests and data between them.
Gateway Aggregation	Combine numerous separate requests into a single request, using a gateway.
Gateway Offloading Pattern	A gateway proxy can be used to offload shared or specialized service functionality.
Gateway Routing Pattern	Use a single endpoint to send requests to numerous services.
Geodes Pattern	Backend services should be distributed among a number of geographical nodes, each of which can handle any client request from any region.
Priority Queue Pattern	Prioritize requests sent to services so that higher-priority requests are received and processed faster than lower-priority requests.
Publisher/Subscriber Pattern	Allow an application to asynchronously announce events to several interested consumers without coupling the senders and recipients.
Queue-Based Load Leveling Pattern	To ease intermittent heavy loads, use a queue that functions as a buffer between a task and the service that it invokes.
Retry Pattern	Allow an application to handle expected, transient errors, while connecting to a service or network resource by transparently retrying a previously unsuccessful operation.
Sequential Convoy Pattern	Process a collection of relevant messages in a predefined sequence, while preventing other groups of messages from being processed.
Sidecar Pattern	Isolate and encapsulate application components by deploying them into a separate process or container.
Static Content Hosting Pattern	Static material should be deployed to a cloud-based storage provider that can deliver it to the client directly.

Strangler Fig Pattern	Progressively replace individual bits of functionality with new applications and services to gradually move a legacy system.
Throttling Pattern	Control the number of resources consumed by an application instance, a single tenant, or an entire service.
Valet Key Pattern	Use a token or key to give clients immediate access to a resource or service that is limited. Data management and security are two of the most important aspects of data management.

We described some different patterns that we can use to implement a cloud-oriented application. In the next section, we will describe the different cloud technologies, but if you want to learn more about Azure cloud design patterns, you can check this book: `https://www.packtpub.com/product/implementing-azure-cloud-design-patterns/9781788393362`.

Using different cloud technologies

In the previous section, we discussed patterns, but now, we will describe the different cloud technologies that we can use to build or deploy any cloud-oriented application.

Containers

A container environment is a common approach to migrating any application and takes advantage of cloud computing and a modern and optimized environment. Some benefits to consider include operating system virtualization, but we don't have the full benefits of the container-based application architecture.

Containerization is a software development technique in which an application or service, along with its dependencies and settings, is packaged as a container image.

Using container technology, you can quickly build, test, and deploy your applications using the same container images.

Containers on a shared OS isolate applications from each other, resulting in a substantially smaller footprint than **virtual machine** (**VM**) images.

Containers and VMs share the same roles – the goal is to isolate an application and its dependencies into a self-contained unit that can run anywhere, taking into consideration that containers and VMs eliminate the need to have physical hardware in order to use computing resources efficiently, both in terms of energy consumption and profitability.

The main difference between containers and VMs is in their architectural approach. A VM is basically an emulation of a real computer that runs programs like a real computer. VMs run on a physical machine using a hypervisor. In the following figure, we can see the difference between a container and a VM:

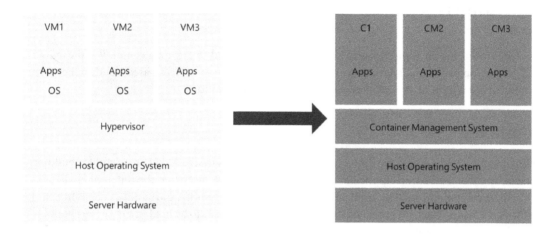

Figure 5.3 – Container versus VM environment

Docker is a tool that facilitates the creation, deployment, and execution of applications using a containerization approach. These containers are lightweight and take less time to boot than traditional servers. These containers also increase performance and reduce costs, while providing appropriate management of resources. Another advantage of Docker is that it is no longer necessary to pre-allocate RAM to each container.

For more details about containers, you can check *Chapter 8*, *Building a Containerized App Using Docker and Azure Container Registry*.

Container orchestration

Container orchestration automates the operational efforts required to run containerized workloads and services.

This includes provisioning, deployment, scaling (up and down), networking, load balancing, and other tasks that software teams must perform to manage a container's lifespan.

Kubernetes is an open source platform. If we have multiple containers to manage, we need to use container orchestration to manage the communication between containers. It allows developers to build containerized applications and services, and we are able to scale, schedule, and monitor those containers easier.

For more details about container orchestration, check *Chapter 9, Understanding Container Orchestration.*

Service Fabric

Service Fabric is a large-scale distributed systems platform that allows you to package, deploy, and manage stateless and stateful distributed applications and containers.

Service Fabric runs on Windows and Linux, on any cloud, and in any data center. It represents a next-generation platform for developers to build and manage microservices-based cloud applications running in containers. It also handles stateful and stateless operations.

Summary

In this chapter, we presented the different cloud design patterns in depth that are useful for building reliable, scalable, and secure applications in the cloud. We first discussed the different challenges in cloud development, such as application security, interoperability, portability, scalability, and performance issues. Next, we presented the different design patterns to be used when we start building a cloud-oriented application, and finally, we presented the different technologies in the cloud, such as containers, orchestration, and Service Fabric.

In the next chapter, we will set up our environment to start discovering the different services to use in the cloud.

Further reading

To learn more about cloud design patterns, you can follow this link: `https://learn.microsoft.com/en-us/azure/architecture/patterns/`.

6

Setting Up an Environment to Build and Deploy Cloud-Based Applications

When creating a web application that does not require a remote development environment, all developers will have their own local settings.

This chapter is intended to help you understand how to prepare your development environment to better manage deployments in your development workflow, by providing some best practices for deployment. We will set up an environment to build and deploy cloud-native applications, and we will define all the prerequisites to implement and deploy any solution.

In this chapter, we're going to cover the following main topics:

- Prerequisites to build a cloud-native application
- Setting up a development environment on Windows
- Setting up a development environment on Linux

Prerequisites to build a cloud-native application

Are you interested in building robust, enterprise-class, cloud-native applications on a large scale? You will need to prepare your development environment before doing so!

In this section, we will describe the different tools, frameworks, and technologies that we can use to build cloud-oriented applications in any environment.

Visual Studio 2022

Visual Studio is an **integrated development environment** (**IDE**) from Microsoft that can be used to develop consoles, **graphical user interfaces** (**GUIs**), Windows forms, web services, and web applications.

Visual Studio has many great features. One of the important features is seamless cloud integration, which makes cloud deployment even easier. It provides all the templates needed for common application types and local emulators. You can provision all dependencies, such as Azure SQL Database and Azure Storage accounts, and you can also quickly diagnose problems with remote debuggers that connect directly to your application.

Visual Studio is available for Windows and macOS. You can download Visual Studio from `https://visualstudio.microsoft.com/`. The Community Edition is free to use and includes a Service Fabric template. You can also use the other editions.

Visual Studio Code

Visual Studio Code is a standalone source code editor. It is the best editor for JavaScript and web developers, with extensions that support almost any programming language.

Microsoft C# for Visual Studio Code, Docker, and Azure App Service extensions must be installed. Visual Studio Code is available for Windows, macOS, and Linux.

You can download Visual Studio Code from `https://code.visualstudio.com`.

Microsoft Azure

Microsoft Azure is Microsoft's public cloud computing platform. It provides various cloud computing services and is made up of data centers from around the world. Each data center is equipped with independent power, cooling, and networking. A region represents a collection of data centers.

Azure offers more global regions than any other cloud provider, with more than 60 regions representing over 140 countries.

A list of regions and their locations is available at `https://azure.microsoft.com/en-us/global-infrastructure/locations/`.

A list of geography locations is available at `https://azure.microsoft.com/en-us/global-infrastructure/geographies/`.

It also includes open source solutions and proprietary technologies from Microsoft and other companies. Originally, the platform was called Windows Azure, but it was renamed Microsoft Azure in 2014. It is huge and grows in size every month. If you are considering moving to Azure, you may feel overwhelmed, since there is so much to explore.

Azure is a subscription-based service. In order to use any Azure service, you will need to have an active subscription. Microsoft offers several choices for getting a subscription; there are options for businesses, the government, nonprofits, students, and individuals.

You can subscribe to Azure at `https://azure.microsoft.com/`.

Azure Cloud Shell

Azure Cloud Shell allows an administrator to run Azure PowerShell or Azure CLI (Bash) commands through the browser. Authentication is automatic.

When you are logged into the Microsoft Azure portal and are on the main portal screen, if you look at the very top of the screen, just to the right of the search bar, you'll notice an icon for Azure Cloud Shell.

Figure 6.1 – Enabling Azure Cloud Shell

After clicking on the Cloud Shell icon, you are prompted to create a storage account, which will be used to store the results of your commands as well as the files you have transferred to Azure. A dedicated Cloud Shell virtual machine is also instantiated. It's temporary, but your files are preserved in two ways – via a disk image and via a mounted file share named `clouddrive`. In the following figure, you can see Cloud Shell:

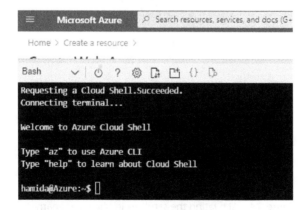

Figure 6.2 – Azure Cloud Shell

You can access Cloud Shell at `https://shell.azure.com/`.

You can switch between **Bash** and **PowerShell**, as shown in the following figure:

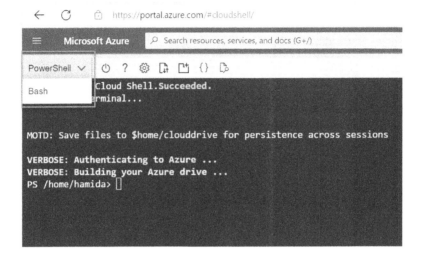

Figure 6.3 – Cloud Shell – Bash or PowerShell

You can use Cloud Shell in Microsoft documentation and Microsoft Learn code snippets using a sandbox environment. Select **Sign in to activate sandbox** to enable the sandbox, as presented in the following figure.

Figure 6.4 – Activating the sandbox in Microsoft Learn

Select **Review permissions** because Microsoft Learn needs your permission to create Azure resources.

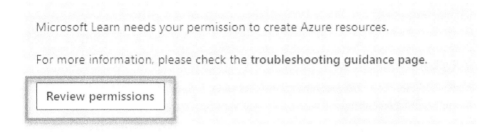

Figure 6.5 – Assigning permissions to the Azure sandbox

You can explore the learn sandbox by following this link:

`https://learn.microsoft.com/en-ca/training/modules/describe-core-architectural-components-of-azure/4-exercise-explore-learn-sandbox`

After the sandbox is enabled in Microsoft Learn, we can select the **Try It** button that is displayed with the Azure PowerShell and Azure CLI code snippets.

Figure 6.6 – Trying code snippets in the sandbox

Azure Cloud Shell is a browser-accessible shell implementation built into many Microsoft tools. Cloud Shell links can be found in Microsoft Docs in the Azure portal. There are also ways to integrate with tools such as Visual Studio Code.

Docker Hub account

To build container images and run containers on our servers, local machines, or virtual machines, we will use Docker.

Docker Hub is an online repository for sharing Docker images. You can push images that you've built up to Docker Hub, and then pull them down at a later time or on a different machine. Storing public images on Docker Hub is free, so it's a useful way to share and maintain images that you use. To use Docker Hub, sign up for an account at `https://hub.docker.com/`, and then in the terminal, run the Docker authentication page and enter your credentials.

Docker Desktop

Docker Desktop is an application for macOS or Windows environments, and it is easy to install. It allows you to build and share microservices with containerized applications.

There have been several evolutions of the Docker Desktop experience over the years, starting with Boot2Docker and going all the way up to the modern tool, Docker Desktop.

Docker Desktop includes the Docker Engine, the Docker CLI client, Docker Compose, Docker Content Trust, Kubernetes, and a credential helper.

Docker Desktop is available for macOS, Linux, and Windows, and you can download it from `https://www.docker.com/get-started/`.

Podman

Podman (**the POD MANager**) is an open source project. It is a daemonless container engine for managing **Open Container Initiative** (**OCI**) images, containers, volumes mounted on these containers, and pods made up of containers. Podman runs containers on Linux but can also be used on Mac and Windows systems with Podman-managed virtual machines. Podman is based on the container life cycle management library libpod and is also included in this repository. The libpod library provides APIs for managing containers, pods, container images, and volumes.

Podman provides a Docker-compatible command-line frontend that uses the Docker CLI aka `docker=podman`. Podman also provides a socket-enabled REST API service that allows remote applications to launch on-demand containers. This REST API also supports the Docker API, so `docker-py` and `docker-compose` users can work with Podman as a service.

Podman Desktop

Podman Desktop leverages the Podman Engine to provide a lightweight, daemonless container management tool. This makes it easy to work with containers from your local environment without having to run container management commands.

Podman Desktop offers all the functionality of Podman, with the added benefit of a powerful and easy-to-use GUI. It's very easy to customize.

You can follow this link to install Podman Desktop: `https://podman-desktop.io/docs/Installation`.

Git

Git is a source code management tool; it helps you manage code in a project as you develop new features, find and fix problems, and simplify collaboration with other developers remotely. You can use Git to manage complex projects, helping you to write better code too. Whether your goal is to manage your own code, collaborate with other developers, or contribute to open source projects, Git is a powerful and helpful tool.

By using Git, you can keep track of the changes you make to files and directories, and it's especially good at keeping track of text changes.

Git is available to download on any operating system (Windows, Mac, and Linux), and you can download and install it from `https://git-scm.com/`.

Azure DevOps

Azure DevOps offers multiple features and services, allowing developers to plan their work using agile methodology, collaborate on their code development in a quick and easy way to improve code production, and implement and deploy modern applications in any language. Azure DevOps has a collaborative culture based on a set of processes, bringing developers, project contributors, and project managers together to build software.

You can create an Azure DevOps account at `https://dev.azure.com`.

GitHub

GitHub is a code-hosting platform for version control and collaboration. It allows you and others to work on projects together wherever you are. In this section, we will learn the basics of GitHub, such as repositories, branches, commits, and pull requests.

You can create an account on GitHub at `https://github.com/`.

Repositories

Repositories are the base building block for GitHub. They can be seen as a folder for your project, and all files that are related to your project will reside in that repository, including the history of your files. When you want to start using GitHub, you'll need to create a repository. It's possible to create a repository through the GitHub interface and clone that to your local machine, using the command-line interface. Alternatively, you can initiate your work locally and push that to a repository on GitHub.

Repositories can be public or private. If a repository is public, it means that its code and basically everything surrounding the repository are publicly available on the GitHub website. Everyone can see the repository, but not everyone can make changes to it directly. In a private repository, only users with access to it can see it and interact with it. Every type of account has the ability to create public or private repos on GitHub. You already saw the differentiation earlier, but I want to repeat it here, since you now already have a better understanding of what GitHub will do for you and your projects. When working with Git locally, we had three areas – the working directory, the staging area, and the local Git repo. The latter is basically Git's internal database that tracks changes. You can work with Git, but most also have a remote control. When in use, GitHub is the remote repository. It will be the place where several users on the team will go to synchronize the changes from their local repositories. Using this remote also makes it possible to share code with others. Although GitHub works in a distributed way, the remote in this scenario can be compared with the central server in traditional source management systems. What you see here is the repository landing page on GitHub. It shows all the files that are included in the project, and it offers many actions that we can do in the repository. We can, from GitHub's interface, create or edit existing files, see the project's README file, clone it to our local machine, or work with its settings.

Branches

The concept of branching is certainly nothing new, and it's also not something that is unique to Git. Other traditional version control systems also offer the ability to create a branch. If you aren't familiar with the concept, it's really nothing complex. Basically, instead of writing all the code on one line and continuing to build on that, we, at some point, diverge from the main development line. See it as a train track that goes on and on and on, and at some point, a sidetrack rises. On that sidetrack, we can do all kinds of things with the code without impacting the main line of code (i.e., the main track). If we like what we have done in the sidetrack, though, we can, at some point, just like train tracks, converge the tracks. The changes we have made in the sidetrack are then merged with what we have in the main development track. Branching is available in other source management systems as well.

Commits

Similar to saving edited files, a commit records changes to one or more files on a branch. Git assigns each commit a unique ID called an SHA or hash. A commit identifies specific changes, when the changes were made, and who made them. When you commit, you should include a commit message that briefly describes your changes.

Pull request

Pull requests track the changes you've pushed to a branch in a repository on GitHub. When a pull request is opened, you are able to review potential changes with collaborators, and you can add follow-up commits before merging the changes into the base branch. In the next section, we will present the different steps for setting up a development environment on Windows for containers and orchestrators.

Setting up a development environment on Windows

In this section, we will set up the development environment on Windows to develop and deploy containers. You can start by installing Visual Studio 2022 or Visual Studio Code.

Afterward, we will configure Docker Desktop for Windows.

Installing and configuring Docker Desktop on Windows

In order to successfully install Docker Desktop, your Windows machine must meet the following requirements:

- **For Windows 10 64-bit**: If you have the Home or Pro version, you need to have Pro 21H1 (build 19043) or higher; otherwise, you need to have the Enterprise version. If you have the Education version, you need to have the 20H2 (build 19042) version or higher.

- **For Windows 11 64-bit**: If you have the Home or Pro version, you need to have 21H2 or higher; otherwise, you need to have the Enterprise version. If you have the Education version, you need to have the 21H2 build version or higher.

You also need to enable the WSL 2 feature on Windows. You can install it using this command line on PowerShell:

```
wsl --install
```

After downloading Docker Desktop, we will install it, and it is easy to configure. Double-click on the executable downloaded file and take a moment to log into your Docker account. If you don't have a Docker ID, it's well worth setting one up. This is what you'll use to push and pull private images, so it is useful beyond just downloading. If you don't have an account already, hit the **Sign-Up** button and go ahead and make one.

Docker needs to activate Hyper-V (the Windows virtualization platform) so that it can create Linux containers in a Windows environment. In the installation window, ensure that the **Use WSL 2 instead of Hyper-V** option is selected on the **Configuration** page, depending on your choice of backend.

To use Docker, we're going to bring up Command Prompt. Run Docker info using PowerShell, and this will print out information about both the client side and server side of Docker. We can observe the result in the following figure:

```
PS C:\WINDOWS\system32> docker info
Client:
 Context:     default
 Debug Mode: false
 Plugins:
  buildx: Docker Buildx (Docker Inc., v0.8.2)
  compose: Docker Compose (Docker Inc., v2.5.1)
  sbom: View the packaged-based Software Bill Of Materials (SBOM) for an image (Anchore Inc., 0.6.0)
  scan: Docker Scan (Docker Inc., v0.17.0)

Server:
 Containers: 0
  Running: 0
  Paused: 0
  Stopped: 0
 Images: 15
 Server Version: 20.10.14
 Storage Driver: windowsfilter
  Windows:
 Logging Driver: json-file
 Plugins:
  Volume: local
  Network: ics internal l2bridge l2tunnel nat null overlay private transparent
  Log: awslogs etwlogs fluentd gcplogs gelf json-file local logentries splunk syslog
 Swarm: inactive
 Default Isolation: hyperv
 Kernel Version: 10.0 22000 (22000.1.amd64fre.co_release.210604-1628)
 Operating System: Windows 10 Pro Version 2009 (OS Build 22000.739)
 OSType: windows
```

Figure 6.7 – Docker info on PowerShell

We can check our local containers and images in Docker Desktop, as shown in the following figure:

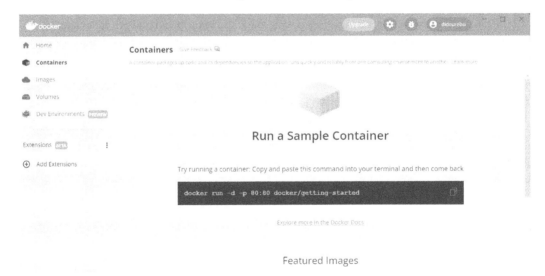

Figure 6.8 – Docker Desktop on Windows

Another way to install Docker Desktop is by using a command line. After downloading `Docker Desktop Installer.exe`, you can run this command line in a terminal:

```
"Docker Desktop Installer.exe" install
```

Alternatively, in PowerShell, you can run this command line:

```
Start-Process '.\win\build\Docker Desktop Installer.exe' -Wait install
```

To start Docker Desktop manually, you can search for `Docker` and select **Docker Desktop** in the search results.

Configuring Git on Windows

We're going to install Git on Windows. The place we visit to get the Git software is the main website: `https://git-scm.com/`. This website includes a lot of information about Git, as well as installers for different operating systems, including Windows.

In the following figure, you can see how to download Git for Windows. First, click on **Download for Windows**, and the **Git – Downloading package** page will come up. Git will automatically start downloading:

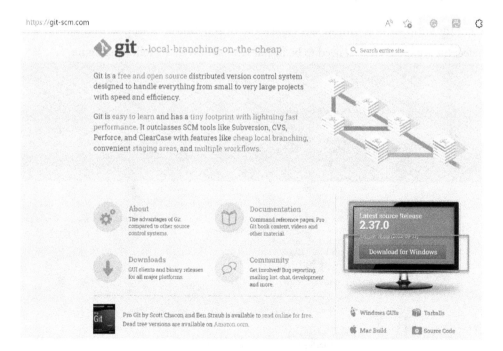

Figure 6.9 – Git for Windows

Once it is downloaded, we can go to the downloaded folder and double-click on it. That'll open up the installer. We can then follow the instructions inside the installer.

Figure 6.10 – The Git setup window

The following are the installation steps:

1. Once the setup script opens, we can choose the **Next** option to accept the license.

2. Select the components that need to be installed, as shown in the following figure:

Figure 6.11 – Git components

3. Select the default editor to be used by Git, as shown in the following figure:

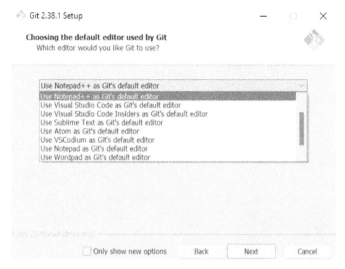

Figure 6.12 – The Git default editor

4. Select **Let Git decide** to allow Git to use its default branch name (which is currently **master**) for the initial branch in newly created repositories. Otherwise, select **Override the default branch name for new repositories** if we need to change the initial branch name.

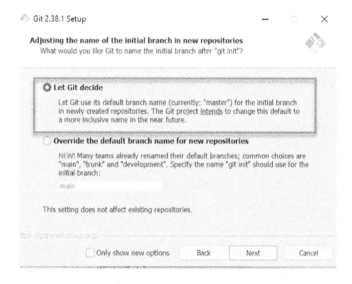

Figure 6.13 – Configuring the initial branch name

5. Adjust the set PATH environment. We can select from different options, including using Git from Bash only. This is the safest choice and your PATH environment will not be modified. The third option consists of using Git and optional Unix tools from Command Prompt.

Figure 6.14 – Configuring the PATH environment

6. Select **Use bundled OpenSSH** to use ssh.exe, or you can use an external OpenSSH.

7. Select **Use the OpenSSL library** in the **Choosing HTTPS transport backend** window.

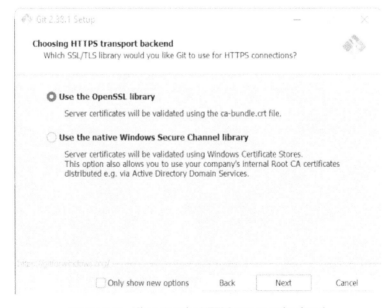

Figure 6.15 – Choosing the HTTPS transport backend

8. In the **Configure the line ending conversions** window, select **Checkout as-is, commit as-is**. As a result, Git will not perform any conversion when checking out or committing text files. This option is not recommended for cross-platform projects. Alternatively, we can select another option, as shown in the following figure:

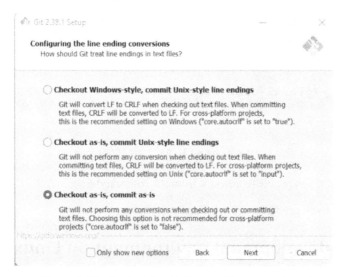

Figure 6.16 – Configuring the line ending conversions

9. In the terminal emulator configuration window, we can select the terminal emulator we want to use with Git Bash. It gives us a choice to either use the MinTTY terminal emulator or Windows' default console. If you're on Windows 10 or 11, go ahead and use the default one. If you're using something older, you might want to give MinTTY a try. Then, click **Next**.

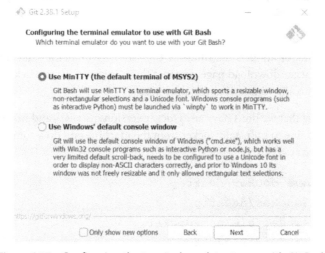

Figure 6.17 – Configuring the terminal emulator to use with Git Bash

We get a couple of extra options after Git is installed, including whether we want to launch Bash or the view release notes.

Now, if you go to the command-line program (that's the console in Windows), you should be able to type `git --version` and get a message saying that you have Git installed. If you're able to type `git --version` and get a meaningful result, execute the following commands from the command line to complete the Git configuration.

Set your username:

```
$ git config --global user.name "FIRST_NAME LAST_NAME"
```

Set your email address:

```
$ git config --global user.email "MY_NAME@example.com"
```

Git is now installed, and you're able to access it from the command line.

In the next section, we will prepare our environment on Linux.

Setting up a development environment on Linux

In this section, we will set up a development environment on Linux. We will start by installing and configuring Docker Desktop on Linux and configuring Git.

Installing and configuring Docker Desktop on Linux

Installing Docker Desktop on Linux is much more straightforward than on other platforms. Since it's just another program on Linux, you can go ahead and install it like any other program. We will run Ubuntu, so we have to search for `Docker on Ubuntu` and head on over to the official Docker installation guide to see how to install Docker on Ubuntu.

If you're running Red Hat or another distribution, the process will be the same, but you'll land on a different page. The Docker download page provides the `.deb` and `.rpm` packages from Ubuntu, Debian, and Fedora Linux distributions and architectures.

We will start by ensuring that we don't have any Docker version installed, and to do that, we can run the following command line to uninstall Docker from the machine and do a complete cleanup, with a purge of the remaining system service files:

```
$ sudo apt remove docker-desktop
$ rm -r $HOME/.docker/desktop
$ sudo rm /usr/local/bin/com.docker.cli
$ sudo apt purge docker-desktop
```

Next, we will install the dependencies that need to be in place before we can add the Docker organization as a source of software for our computer. We want to install the HTTPS Transport, a few certificate authorities, `curl`, and a collection of generally useful software, `software-properties-common`.

In the next step, we will inform the system that it should trust software provided by the Docker organization. This will fetch the cryptographic signing key of the Docker organization and tell our computer to import this as a trusted key to provide software to our system. The system will download the key, and then it will take the contents of that key and pipe them to the `apt-key` command, which adds a trusted source. Then, the next important step is to add the Docker organization as a source of software for this computer.

Regardless of your choice of distribution, you'll need a 64-bit installation and a 3.10 or newer kernel. Kernels older than 3.10 do not have the features Docker needs to run containers. Data loss and kernel panics are common under certain conditions.

We will add an official source of software, which means that we're going to add a source of Debian packages for the `amd64` architecture.

We will call `apt-add-repository`, which adds external repositories using key servers to an APT installation's list of trusted sources. We will use the `lsb_release` command inside a subshell. We will refresh the dependencies again using `apt-get update`.

To install Docker, we will update the existing list of packages using this command line:

```
$ sudo apt-get update
```

Then, we will install the prerequisite packages that allow `apt` to use packages over HTTPS:

```
$ sudo apt-get install \
    ca-certificates \
    curl \
    gnupg \
    lsb-release
```

The gpg key for the Docker repository has to be added to the system, as mentioned in the following command line:

```
$ sudo mkdir -p /etc/apt/keyrings
$ curl -fsSL https://download.docker.com/linux/ubuntu/gpg |
sudo gpg --dearmor -o /etc/apt/keyrings/docker.gpg
```

We will set up the repository by using this command line:

```
$ echo \
  "deb [arch=$(dpkg --print-architecture) signed-by=/etc/apt/
keyrings/docker.gpg] https://download.docker.com/linux/ubuntu \
  $(lsb_release -cs) stable" | sudo tee /etc/apt/sources.
list.d/docker.list > /dev/null
```

We will add the Docker repository to the apt sources:

```
$ sudo add-apt-repository "deb [arch=amd64] https://download.
docker.com/linux/ubuntu bionic stable"
```

Next, we will update the package database with the Docker packages from the newly added repository:

```
$ sudo apt-get update
```

Then, we will install Docker Engine:

```
$ sudo apt-get install docker-ce docker-ce-cli containerd.io
docker-compose-plugin
```

We will check whether we installed the package from the Docker repository instead of the default Ubuntu repository using this command line:

```
$ sudo apt-cache policy docker-ce
```

If docker-ce is not installed, we will use this command line:

```
$ sudo apt install docker-ce
```

By default, Docker commands can only be run by the root user or a user in the Docker group automatically created during the Docker installation process.

We will add a user to the Docker group using this command line:

```
$ sudo usermod -aG docker username
```

Now, you can test whether it worked by typing docker run hello-world. You can add a user to the list of users who are allowed to run Docker without putting sudo at the beginning. Docker will start up a container and run a process called hello-world that introduces you to Docker, and then it's done. It's working!

We will install Docker after downloading the latest `.deb` package using this command line:

```
$ sudo apt-get install ./docker-desktop-<version>-<arch>.deb
```

To start Docker Desktop for Linux, you can type this command line:

```
$ systemctl --user start docker-desktop
```

When Docker Desktop launches, it creates a dedicated context targeted by the Docker CLI and sets it as the context currently in use.

Configuring Git on Linux

We will download Git from `https://git-scm.com/download/linux` and follow the steps mentioned in the Git documentation. You just type the command line according to the Linux distribution, and then the package manager should go out and get the latest version of Git and install it. After you have it installed, you can go to the command line and check the Git version. We will use Ubuntu and start by setting up the repository using this command line:

```
$ sudo add-apt-repository ppa:git-core/ppa # apt update; apt
install git
```

Afterward, we will use this command line to install Git:

```
$ sudo apt-get install git
```

We have configured Docker and Git to build cloud-oriented applications using Visual Studio 2022 or Visual Studio Code.

Summary

In this chapter, we prepared a development environment to build and deploy cloud-oriented applications, and we installed and configured Docker Desktop and Git on Windows and Linux.

In the next chapter, we will deploy a web application on Azure App Service using the Azure portal, Visual Studio, and Azure CLI commands. We will describe Azure App Service's key components and values, and we will cover the management of authentication and authorization during deployment.

Questions

To set up a development environment on Windows, what are the necessary tools to install to build and deploy a cloud application? What about for a Linux environment?

7

Using Azure App Service to Deploy Your First Application

In this chapter, we will deploy an ASP.NET Core MVC web application using Azure App Service. We will cover the creation, updating, and scaling of an app in Azure App Service. We will also describe the Azure App Service key components and their value, and the different ways to configure web app settings. We will see how to deploy an app using Azure App Service using the Azure portal, Visual Studio, or Azure CLI commands, and we will cover the management of authentication, authorization, and deployment.

In this chapter, we're going to cover the following main topics:

- Azure App Service basics
- Creating and configuring web app settings in Azure App Service
- Deploying an application on Azure App Service
- Scaling apps in Azure App Service
- Exploring Azure App Service deployment slots

There are several ways to move an application to Azure. A common method is to migrate your application to Azure App Service. It is a managed service that can build and execute a web application in the cloud. In some cases, you may need to change or redesign your application.

Azure App Service is an HTTP-based service to host web applications such as REST APIs, web services, **Windows Communication Foundation** (**WCF**), and mobile backends. You can use it for executing and scaling applications in Windows- and Linux-based environments. Several languages are supported, such as .NET, .NET Core, Java, Ruby, Node.js, PHP, and Python.

Azure App Service basics

In this section, we will discover Azure App Service's basic components and the different features for building and deploying an application.

Azure App Service is a platform-as-a-service offering with a ready-to-use runtime environment. We can use the Azure portal to create and configure the service or run PowerShell scripts or other executables as background services.

It's great for handling legacy apps or parts of your apps that can't be immediately migrated to Azure App Service, and when you're ready to deploy, you can do so in a Windows or Linux environment. By building and hosting your applications with Azure App Service, you can take advantage of so many services within Azure to make them better. One of the best services to use with Azure App Service is Azure DevOps.

Azure App Service conforms to the ISO, SOC, and PCI standards by default. You can also set up IP restrictions, configure subdomains, and create blacklists and whitelists, among other compliance-focused requirements. As an extra convenience, Azure App Service integrates natively with Azure Active Directory so you don't have to write any authentication mechanisms from scratch.

When we host an application on Azure App Service, we can manage the scale through the Azure CLI or the Azure portal. You can scale up or out, depending on the application and your needs during a specific period. For example, when we want to scale up, we can increase the CPU usage, memory, or disk space.

Before creating a new resource Azure App Service, we need to understand more about Azure App Service plans. In addition to App Service plans, there's a concept called an **App Service Environment** (**ASE**). This is where we can run isolated web applications. So, we'll go ahead and get a better understanding of these concepts, and then we will walk through creating web apps in the portal and using the Azure CLI and PowerShell.

Azure App Service plans

Depending on the type of compute environment you use, the App Service plan environment you select dictates the costs. When we create a new web app, we need to select the App Service plan, as depicted in the following figure:

Create Web App   ...

App Service Plan

App Service plan pricing tier determines the location, features, cost and compute resources associated with your app.
Learn more ☐

Linux Plan (Central US) * ⓘ   (New) ASP-packtrg-a1ba   ⌄
                                      Create new

Sku and size *   **Premium V2 P1v2**
                                      210 total ACU, 3.5 GB memory
                                      Change size

Figure 7.1 – App Service plan for a Linux environment

The App Service plan is used to define where the resources are going to be created, the number of virtual machine instances, and the number of virtual machines. When selecting your App Service plan, there are several tiers that you can choose from.

We can group them into two categories:

- Non-isolated

- Isolated

You can see in the next figure the different categories we find in the App Service plan section when we create a new Azure App Service resource. With the **SKU and size** element, we can click on **Change size** to update the default one:

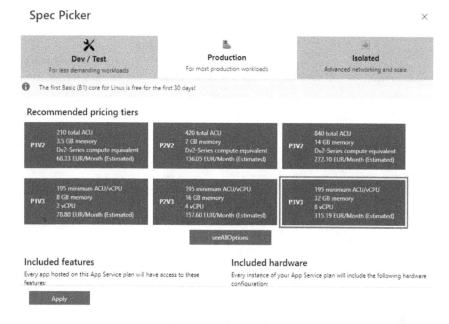

Figure 7.2 – SKU and size

Let's explore these categories in detail, and we will present the subcategories for **SKU and size**:

- **Non-isolated**: This can be a shared computer or dedicated compute resource. The difference between them is that with a dedicated compute resource, the infrastructure is shared; it is another non-production-grade type of Azure App Service plan. We can use these tiers for testing and development and light workloads. The lowest tier, F1, is free for 60 minutes per day. If you need your resources to scale, then you need to move up to the next tier, which is a dedicated compute resource. This tier is for most workloads and supports custom domains, **Secure Socket Layer** (**SSL**), autoscaling, staging slots, backups, and the Traffic Manager service. There are a few categories of pricing tiers:

  - **Free or Shared (F1 or D1)** – In the following figure, we can select them in the **Dev / Test** tab:

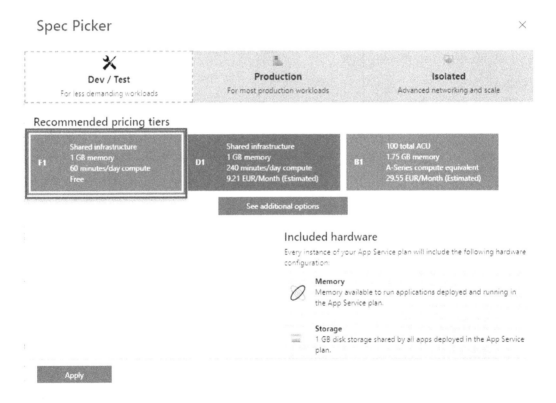

Figure 7.3 – Free or Shared (F1 or D1) plan

This plan is used to test Azure App Service resources but isn't used for production purposes.

- The **Basic (B1, B2, or B3)** pricing tier is designed for applications with low traffic requirements:

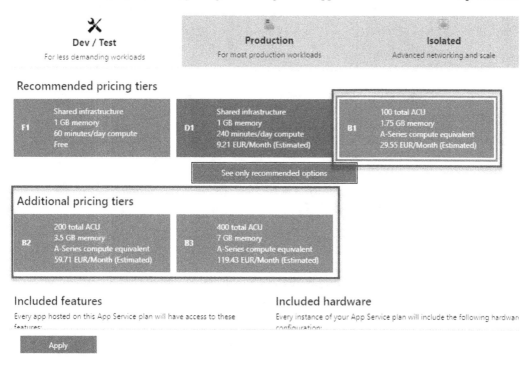

Figure 7.4 – Basic plan in Azure App Service (B1, B2, or B3)

- You don't need advanced autoscaling or any traffic management features – these can be used for testing applications, for example:

  - **Standard (S1, S2, or S3)** is designed for running production workloads, and this is based on the number of instances

  - **Premium v2 (P1v2, P2v2, or P3v2)** is designed to provide enhanced performance for the application's production environment

  - **Premium v3 (P1v3, P2v3, or P3v3)** is just an upgrade of the Premium plan

In the following figure, you can select the **Production** tab in order to select the Standard or Premium plan:

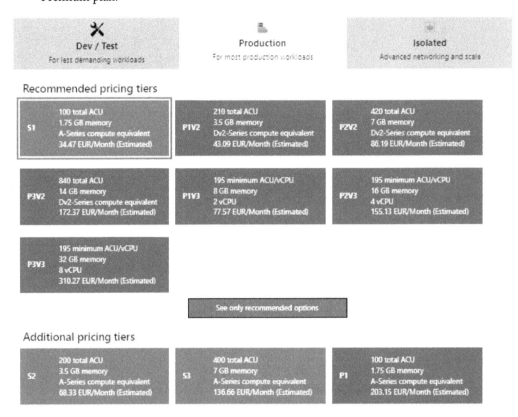

Figure 7.5 – Production tab showing App Service plans

To see more details about these plans for Windows, you can follow this link: `https://azure.microsoft.com/en-us/pricing/details/app-service/windows/`. For Linux, you can follow this link: `https://azure.microsoft.com/en-us/pricing/details/app-service/linux/`.

- **Isolated**: You would select any of the plans within this tier if you needed an isolated network or more control, as it's a single-tenant system and it supports private app access. This is known as an **ASE**. An ASE is a completely isolated environment dedicated to the execution of your web applications. It's still part of App Service, and it's still a managed service, but the big difference here is you get the entire compute environment dedicated to you, and you also get the isolation of being able to use it in your own VNet. It is designed for high scale and high memory isolation. This tier will scale up to 100 instances.

Another benefit of an Isolated plan is that when you are working in a hybrid environment, the apps running in an ASE can take advantage of existing on-premises VPN connections. You might have a really fast connection back to the on-premise environment, and this might be a really interesting way to allow on-premise applications to communicate with apps that you've got running on your own VNet in an ASE. An ASE is a deployment of Azure App Service in its own VNet based on the new capabilities of ASEs, running in a separate **stock-keeping unit** (**SKU**) called an isolated SKU.

We will look at that in the next section.

App Service environments

An ASE is a dedicated environment that runs Azure App Service, and we can run it at a high scale of up to 100 instances. We can still run workers on Windows or Linux. We have worker sizes that are a little bit bigger in terms of memory (RAM), so the ratio of CPU and RAM is more heavily weighted on the RAM side. An important feature of an ASE is that it allows network isolation. This means that if we deploy any ASE on an Azure VNet, we are able to lock down that VNet in order to not block the internet access, and we can still run private web applications on that VNet.

You can see more about App Service plans at this link: `https://azure.microsoft.com/en-ca/pricing/details/app-service/windows/`.

Next, we will create an App Service plan to explore the different configurations.

Creating and configuring an App Service plan

Open the Azure portal and select an existing resource group (or create a new resource group). We are going to select **Add**, and then we will search for an App Service plan.

In the following figure, we can see that we don't need to add much information to create the App Service plan. We need to fill in **Subscription** and **Resource Group**, **Name**, select an option for **Operating System**, and select an option from the **Region** dropdown. Next, we can make a **Pricing Tier** selection:

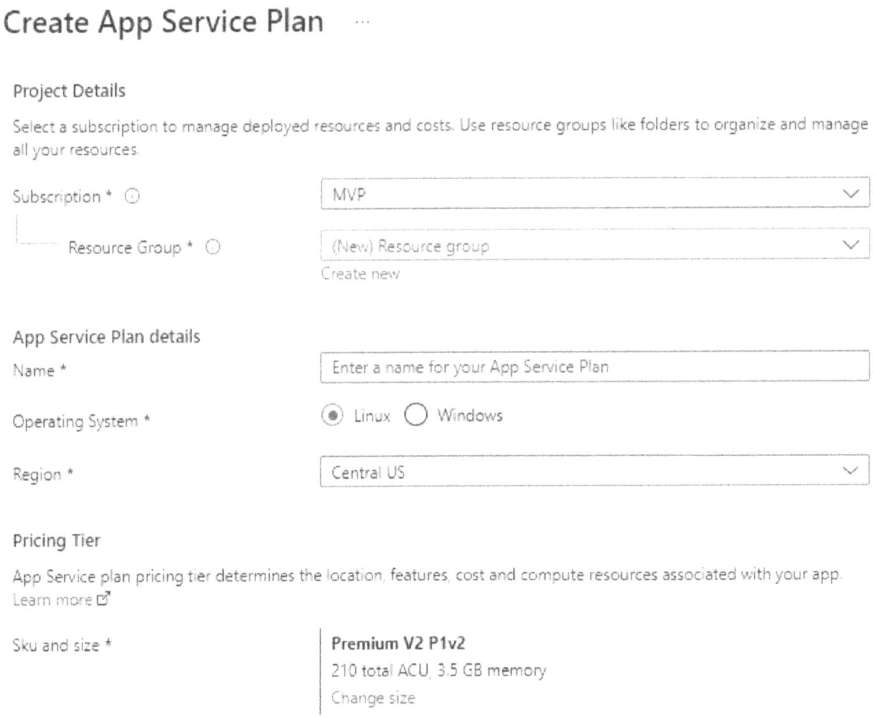

Create App Service Plan ...

Project Details

Select a subscription to manage deployed resources and costs. Use resource groups like folders to organize and manage all your resources.

| Subscription * ⓘ | MVP ⌄ |

| Resource Group * ○ | (New) Resource group ⌄ |
| | Create new |

App Service Plan details

| Name * | Enter a name for your App Service Plan |

| Operating System * | ⦿ Linux   ○ Windows |

| Region * | Central US ⌄ |

Pricing Tier

App Service plan pricing tier determines the location, features, cost and compute resources associated with your app. Learn more ⧉

Sku and size *	**Premium V2 P1v2**
	210 total ACU, 3.5 GB memory
	Change size

Figure 7.6 – Create App Service Plan

We can change the size of the pricing tier by selecting **Change size**, and you'll notice we're in the Production tier. If we scroll down, we can see what is included in every plan.

At the bottom of the page, as presented in the following figure, we can enable **Zone redundancy**, and we can also make an App Service plan's zone redundant after being deployed. You can see more about that at this link: https://docs.microsoft.com/en-us/azure/availability-zones/migrate-app-service.

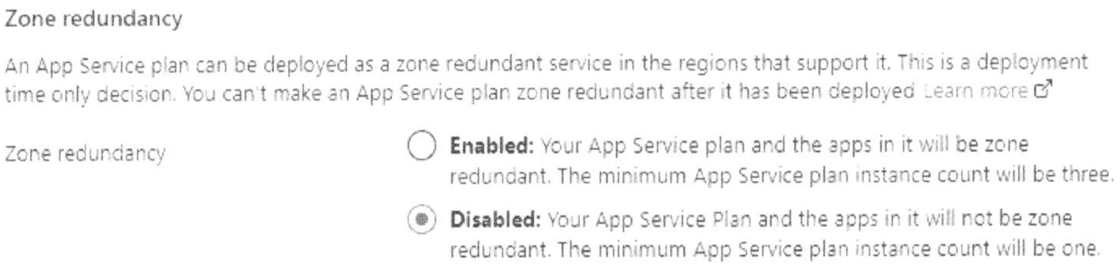

Zone redundancy

An App Service plan can be deployed as a zone redundant service in the regions that support it. This is a deployment time only decision. You can't make an App Service plan zone redundant after it has been deployed. Learn more ⧉

| Zone redundancy | ○ **Enabled:** Your App Service plan and the apps in it will be zone redundant. The minimum App Service plan instance count will be three. |
| | ⦿ **Disabled:** Your App Service Plan and the apps in it will not be zone redundant. The minimum App Service plan instance count will be one. |

Figure 7.7 – Enable or disable Zone redundancy

We will click on **Review + Create** when the validation passes successfully. When we click on the **Create** button, it takes a few seconds for the App Service resource to be created. Once the App Service plan is deployed, we will go to the resource App Service and explore the different elements.

In the **Overview** blade, we can see our App Service plan's name, the tier, and the fact that this instance is a shared instance. Any apps that we have will be listed under **Settings and Apps**. If we wanted to scale out, then we would select **Scale out**, but you will notice that autoscaling is not available for this resource if you have selected the **Free** or **Basic** plan.

In order to have autoscaling available, we have to move up a tier. To do that, it is really simple – you can scale up to move into production, and if you scroll down a little bit, you can see what is supported. You have two options for scaling:

- You can do an instance count, which is manual. For example, we can change our instance count to **3**. Be sure to click on the **Save** button. To verify that we're scaling out, let's check that in **Overview**. We can now see our plan has three instances.
- We can also scale based on a schedule or metrics.

The first thing we're going to do is to scale based on a metric. We don't have a rule so let's go ahead and create one by selecting **Add rule**. You can define your criteria. You need to select your time aggregation and then your metric's name. We're going to leave this as **Average CPU percentage**. Then, you need to scroll down and select your operator. I'm going to leave it at **greater than**. The metric to trigger this action will be 80%.

And now we need to define the duration. For example, we can set the duration as 5 minutes. When our CPU threshold is over 80% for 5 minutes, an action will take place. In this case, we're going to increase the count by two.

Our last setting will be the cooldown. This is the amount of time from when the initial scaling occurred until it will scale again. We can combine scaling with different metrics if we want to; we would just add another rule. Now, you can go ahead and click **Add**. Not only can we use multiple rules for scaling, but we can also scale based on a condition.

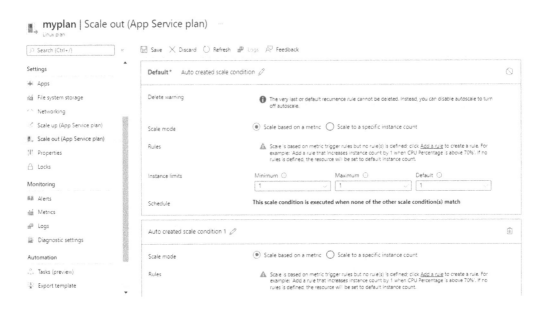

Figure 7.8 – Scale out (App Service plan)

Having created the App Service plan, we will now create a new app using App Service.

Creating a new App Service resource in the portal

Now we can add an App Service resource to the App Service plan just created. Note that this doesn't have to be a two-step process, because you can add the App Service plan where you will create your App Service resource. I'm going to go ahead and select **Add** from the resource group and then search for Web App:

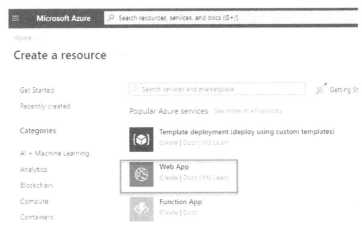

Figure 7.9 – Create a resource – Web App

We don't have to provide a lot of information here. Select your subscription and resource group. We need to provide a web app name under the `.azurewebsites.net` namespace. Note that this must be unique. Later on, you can add a custom domain for your web application. By default, we're going to start with `.azurewebsites.net`. Choose how you're going to publish to this app, either with code or a Docker container. So, for example, when we have code selected here, we can pick a runtime stack such as .NET 6 or 7, .NET Core, ASP.NET, or Java – there's a bunch more on the list. However, if we've already got our application code packaged up in a Docker container, we can just pick **Docker Container** and then decide on the OS environment, which can be Linux or Windows, and that's all we need to do. If we select **Docker Container**, a new tab called **Docker** will appear. We need to select an image source, which can be Azure Container Registry, Docker Hub, a private registry, or a QuickStart sample. We need to select from different options.

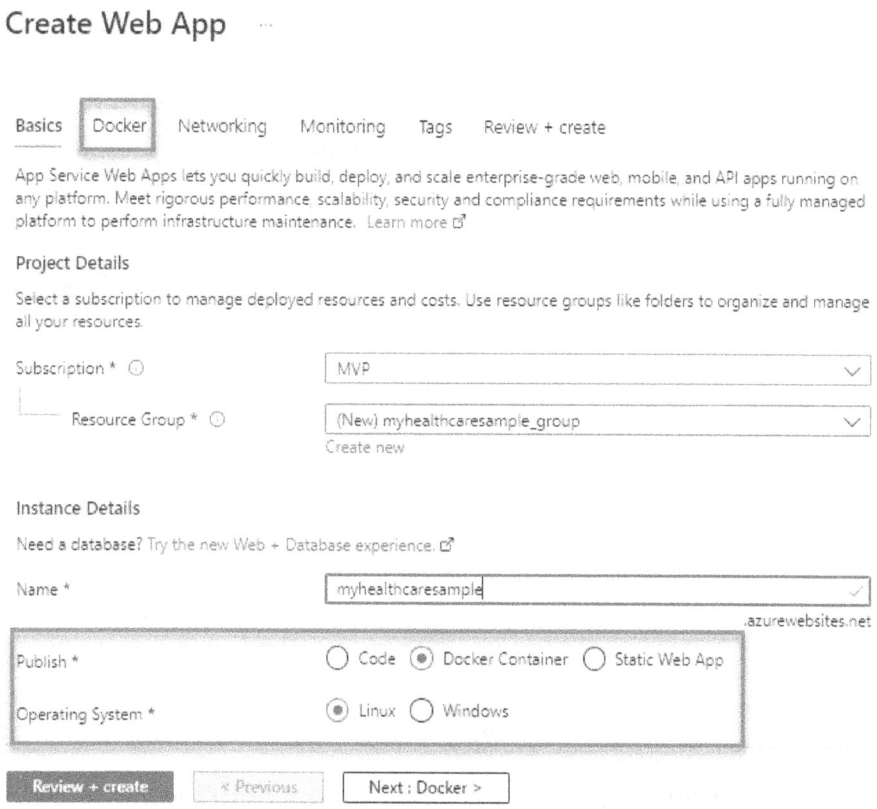

Figure 7.10 – Selecting Docker Container in App Service

We are going to start with code and then we will select the runtime stack, which will be .NET 6 or .NET 7. My operating system will be **Linux**. Next is **Region**, and right under the dropdown for **Region**, it states, **Not finding your App Service Plan? Try a different region** or select your App Service Environment:

Figure 7.11 – Regions in App Service

If we try to select the App Service plan that we created, it's not available. We have to create a new one. Therefore, you may have to change your region. And now we can select that App Service plan. We only have one in the region. At this point, you can go ahead and create a new App Service plan and do everything in one step.

We will keep the default settings for monitoring and disable **Application Insights**. We will then select **Review + Create**. Now, your web app has been deployed, so let's go ahead and take a look at it in the following figure:

Figure 7.12 – App Service Overview

Under **Status**, we can see that our app is up and running. We have the URL for the app and we can see the App Service plan that's associated with this app. We can see how many instances we have. In our case, it is only one instance.

Note that you don't have to create your web app using the portal; you can use your favorite tool, such as Visual Studio, and then push that out to Azure, or use the Azure CLI.

We will see in the next section how to create an App Service resource using the Azure CLI.

Creating a new Azure App Service using the Azure CLI

We built an App Service resource in the portal, so now let's do the same thing using the CLI.

We'll open Azure Cloud Shell from the icon in the portal or by using this link: `https://shell.azure.com`. Or, you can install the Azure CLI on your computer, and then you can authenticate on the Azure platform. Here, we will use the browser. We will use the `az` command, and then we will use subcommands to work with different services.

To host Azure App Service, we need to create a new resource group:

```
$ az group create -n packtrg -l westus2
```

Next, we have to create a plan. We can do this by using the `az appservice plan create` command:

```
$ az appservice plan create --name myserviceplan --resource-
group packtrg --sku D1 --is-linux
```

We will press *Enter* and the App Service plan will be created. The JSON output including all information related to the new App Service plan will be displayed.

We will now create our App Service resource using the plan:

```
$ az webapp create --name myhealthcarewebapp --plan
myserviceplan  --resource-group packtrg
```

You can add the `--deployment-local-git` parameter at the end if you use Git to deploy your application. We will press *Enter* and we will have JSON output.

Note that we could have used different input parameters to create an ASE. You can add more parameters to create a web app. Check this link for more details: `https://docs.microsoft.com/en-us/cli/azure/webapp/config?view=azure-cli-latest`.

Let's move on to the next section, where we will learn how to do all this using PowerShell.

Creating a new App Service using PowerShell

We built an App Service resource from the portal and used the CLI (the Azure CLI); now we will use the same browser to create an App Service using PowerShell, but you can also install Azure PowerShell on your computer. We will follow all the same previous steps: we will create a resource group and an App Service plan, and at the end, our App Service.

To create a new resource group, we will use the `New-AzResourceGroup` cmdlet:

```
PS /home/username> New-AzResourceGroup -Name packtrg -Location
"South Central US"
```

We will use the `New-AzAppServicePlan` cmdlet to build the App Service plan in the `S1` tier:

```
PS /home/username> New-AzAppServicePlan -Name myserviceplan
-Location "South Central US" -ResourceGroupName packtrg -Tier S1
```

And for the last step, we'll use the `New-AzWebApp` cmdlets, and we'll just build the web app using `AppServicePlan` and the resource group that we already built:

```
PS /home/username> New-AzWebApp -Name
myhealthcarewebapp    -Location "South Central US"
-AppServicePlan "myserviceplan" -ResourceGroupName packtrg
```

Now, when it comes to building web applications with Infrastructure as Code, the native solution on Azure is using **Azure Resource Manager** (**ARM**) templates. You can see the different templates that we can use to create a new Azure resource in Azure QuickStart Center (`https://learn.microsoft.com/en-us/azure/azure-resource-manager/templates/overview`). For example, for App Service in Linux, we can follow the previous link and just fill in a few fields for WebAppName, SKU, Runtime Stack, and then the location for the resources.

We created an App Service by using the Azure portal, the Azure CLI, and Azure PowerShell. In the next section, we will create and configure web app settings.

Creating and configuring web app settings in Azure App Service

In App Service, app settings are variables that we pass to the application code as environment variables. In a Linux environment or in custom containers, App Service uses the `--env` flag to pass app settings to the container and set the environment variables.

In both cases, they will be injected into your application environment when the application is launched. In the case of additions, deletions, or modification of application settings, App Service will trigger an application restart.

Application settings can be configured and accessed by navigating to App Service. Under **Configuration**, we select **Application settings**, as presented in the following figure:

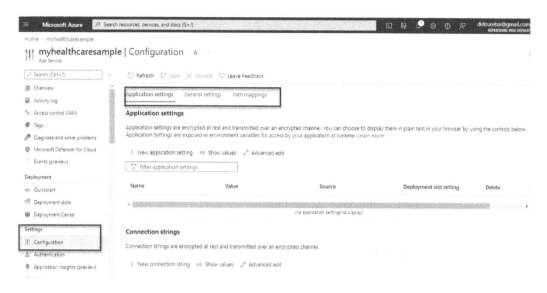

Figure 7.13 – Configuration settings in App Service

When we create an ASP.NET Core application, we have the application settings in `Web.config` included in `<appSettings>` or `appsettings.json`. They are the same as the application settings in App Service. These settings can be the connection string of a database. In Azure, we use them to store credentials such as database passwords because they are always encrypted when stored. When you are in the development environment, the code uses the settings stored locally in `Web.config` or `appsettings.json`, and the same code will use your production secrets when your application is deployed to Azure.

How we can add and edit settings in Azure? Let's look at that next.

Managing settings in Azure

To add a new application setting in App Service, we go back to the previous figure, select **+ New application setting**, and a new window will be displayed to add or edit the application settings. We can add a name with a value, as presented in the following figure. If we use deployment slots (you can learn more about deployment slots in the *Exploring Azure App Service deployment slots* section), we need to specify whether the setting can be swapped or not. Otherwise, if you use only one slot, one slot presents the default production environment, and you fill in the settings in the current slot:

Figure 7.14 – Add/Edit application setting in App Service

Once created, the value will be hidden. If you want to edit a specific application setting, you can select the setting and click on **Edit**, as shown in the next figure:

Figure 7.15 – Edit an application setting in App Service

We can add a connection string to secure the database access in the same tab, **Application settings**. At the bottom, we have **Connection strings**. If we click on + **New connection string**, we then have a new window to fill in all the information required, such as the name, the value, and the type of our database, as shown in the following figure:

Figure 7.16 – Add/Edit connection string in App Service

We can update the settings related to our framework or the platform, for example, in **General settings**. In **Path mappings**, we can configure handler mappings and virtual application and directory mappings.

Having completed the necessary configurations for an application, we will see how we can deploy an application using several methods.

Deploying an application on Azure App Service

Having created and configured an App Service, we will deploy the application on Azure App Service. We will use Visual Studio 2022 to publish the application to App Service.

Open Visual Studio 2022. We need to compile our application in Release mode. We right-click and select **Publish**, and then we will see the following window:

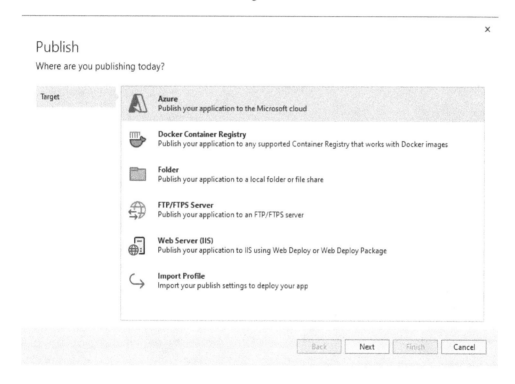

Figure 7.17 – The Publish window in Visual Studio 2022

Select **Azure**. After we click on the **Next** button, we need to select an option from the following window:

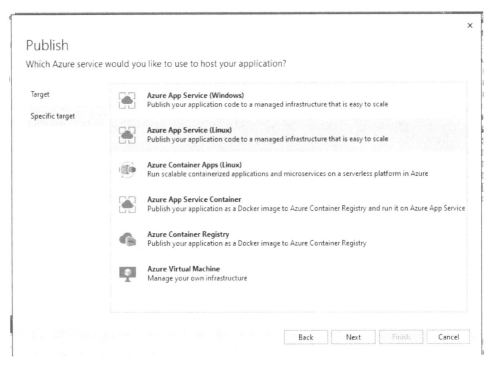

Figure 7.18 – Publishing target in Visual Studio 2022

We created an App Service in a Linux environment, so we will select the **Azure App Service (Linux)** option to publish the application. After, we will click on the **Next** button to select an existing App Service instance. We can also create a new Azure App Service instance in the following window. Note that we have only one deployment slot. By default, it is the production environment, but we can publish to one or more deployment slots (you will learn more about deployment slots in the *Exploring Azure App Service deployment slots* section). We will select the App Service instance name, and then we will click on **Next**:

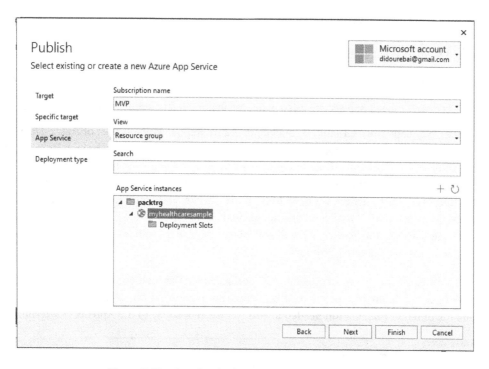

Figure 7.19 – App Service instances in Visual Studio 2022

In *Figure 7.20*, we can select the deployment type. We can publish it and generate a pubxm file, or we can use CI/CD using GitHub Actions workflows:

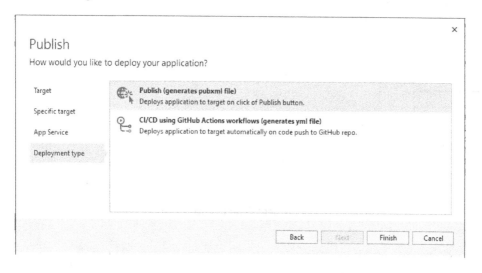

Figure 7.20 – Deployment type in Visual Studio 2022

There are many third-party systems and tools that you could use to deploy code to Azure. We used Visual Studio 2022, which is a simple way to deploy any application. You could use also Visual Studio Code. Visual Studio Code has multiple features:

- It is an interactive tool.

- We can deploy code to App Service using native capabilities by configuring continuous deployment.

- We can pull code from GitHub, Bitbucket, or Azure Repos, which is part of Azure DevOps. We can associate our web app to point to these repositories – a specific branch in these Git repositories – and anytime those branches are updated, it will deploy the update to your App Service web application. This is the continuous deployment option. If you wanted to opt for continuous delivery, where we need to do more checks, you could use Azure DevOps, for example, or any CI/CD tool. So, continuous deployment means that any change in the code will be updated in App Service. We will see how we can use the native continuous deployment model in App Service to deploy code.

We will open the App Service resource again in the portal. Under **Deployment**, we select **Deployment Center**. We've got to define the source control. Where are we storing our code? We will select the source and the different information needed to connect the repository to the App Service resource:

Figure 7.21 – Deployment Center in Azure App Service

We have some options to select as the source:

- **Continuous Deployment (CI/CD)**: **GitHub**, **Bitbucket**, **Local Git**, or **Azure Repos**

- **Manual Deployment (Push)**: **External Git**

We will select a repository in GitHub. We will authorize access to the account. Once connected, we will select an organization, repository, and branch:

Figure 7.22 – Deployment Center – GitHub repository

We will select **Save** to save the different settings. But for a GitHub source, we can change the provider from the default value, **GitHub Actions**, to **App Service Build Service** or **Azure Pipelines**.

If we select **Azure Pipelines**, we have a different setting to set up by selecting **Source** and the credentials. We'll select **Get started**:

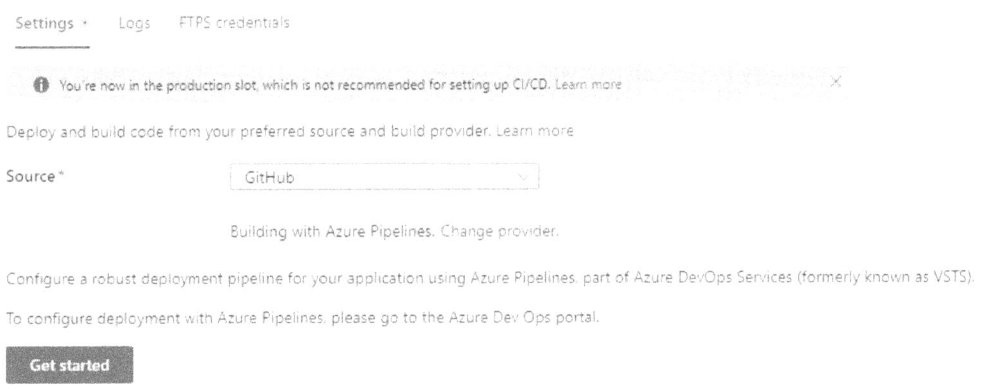

Figure 7.23 – Deployment Center – GitHub source with Azure Pipelines

We've learned how to deploy Azure App Service and deployment slots. In the next section, we're going to see how we scale these application resources in production.

Scaling apps in Azure App Service

Autoscaling allows the system to adjust the resources needed to meet the needs of different users while controlling the costs associated with those resources. Autoscaling is available for many Azure services, including web applications. Autoscaling requires the configuration of autoscaling rules that specify the conditions under which resources are added or removed.

When we run an application in a production environment, the first thing to think about is scalability. We need to understand the difference between scaling up vertically and scaling out horizontally.

If we run a virtual machine using the *Standard_A1_v2* App Service plan (which means that we have 1 CPU core and 2 GB of memory and over time, the application demands more resources), we can easily just go to the properties of that virtual machine or make an API call to make the virtual machine bigger. We can change it to a different size, for example, changing to the *Standard_A8m_v2* App Service plan. In that case, we would have 8 CPU cores and 64 GB of RAM, and it takes just a short outage to take the machine from one size to another. This is the concept of scaling up vertically.

We can change the plan type for an App Service if the application deployed requires more resources. The production-grade App Service plan can support multiple virtual machines. We can scale out horizontally, in this case, if we need to add more than one VM.

Scaling App Service manually

Let's go back to the Azure portal and open our App Service. Under **Settings**, on the left, we've got **Scale up (App Service plan)** and **Scale out (App Service plan)**:

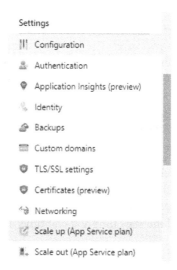

Figure 7.24 – Settings – Scale up and Scale out in App Service

If we need to make the App Service plan bigger, we basically need to increase the number of compute instances (we have a single one now). For example, if we selected the S1 plan, on this scale-up screen, we could take it to the Premium tier. Then, you can scroll down and see more options. Now, when it comes to scaling out, this is where we can scale based on performance, and it's also where we can manually scale the compute instance.

Notice that, right now, it's set as **Manual scale**, **Maintain a fixed instance count**. By taking a look at the following screenshot, you can see that **Instance count** is currently set to just **1**:

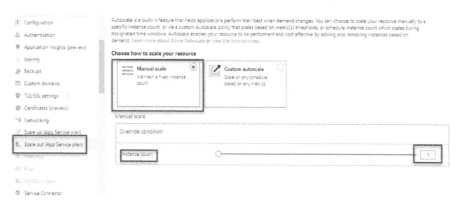

Figure 7.25 – Manual scale in App Service

If we wanted to manually scale this, we could just change **Instance count** to **2**, and then we would just need to click on **Save**. That would manually scale up the instance count. An implicit built-in load balancer will be added and will distribute the connections between the two VMs running this App Service plan.

Scaling based on a schedule

When you have a predictable application, you know exactly what the resource requirements are going to be, scaling on a schedule. For example, for an e-commerce application, on sale days, the application requires more resources. So, we go back to the properties of App Service, and in the **Scale out** option in *Figure 7.25*, we will change it to **Custom autoscale** instead of **Manual scale** so we can scale on a schedule. When we get onto *Figure 7.25*, what we're going to do is delete the default autoscale rule. You can add more than one autoscale condition.

To add a new condition, we will click on + **Add a scale condition**, and then we can add specific conditions, such as multiple specific days of redundancy, or we can select specific start and end dates to add instances. For example, if we select Monday, Tuesday, and Sunday and add an instance, then the instance count will be **2**. We add the start time and end time, and the result is presented in the following figure:

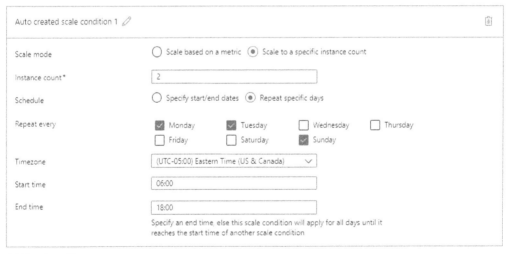

Figure 7.26 – Add a scale condition in App Service

Monitoring autoscaling activity

You can track when autoscaling happened using the **Run history** tag, where we have a graphic presentation of the App Service. The graphic presentation displays the number of instances that have changed over time and the autoscaling conditions that contributed to each change.

The following figure shows an example of monitoring autoscaling in App Service:

Figure 7.27 – Monitoring autoscaling in App Service

And since App Service is a managed service, this makes the environment secure. There's basically a built-in load balancer. Azure Load Balancer is a low-latency Layer-4 load-balancing service with high performance, build to handle a large volume of requests per second in order to ensure the high availability of the solution.

We have the ability to automate the deployment of the code for our application or for the deployment of the infrastructure only. We will explore more details about deployment in the next section.

Exploring Azure App Service deployment slots

Azure App Service supports continuous deployment. The code that is pushed to source control will then be pushed out to the web app automatically. With deployment slots, we can validate changes before pushing the code to production and, of course, we can revert to the previous version if there's a problem. The App Service plan defines the number of deployment slots that are available to you. In the Free, Shared, or Basic plans, you don't have any deployment slots. When you move up to a Premium or Isolated plan, you have up to 20 deployment slots. Each deployment slot has its own hostname, which allows for testing. When you deploy an app to a deployment slot, it's warmed up and this eliminates the downtime when the app is swapped into the production environment.

We can also clone instances within deployment slots. When we clone an instance or an app, the app settings, connection strings, language, framework versions, web sockets, and the HTTP version will be cloned. We can use the **Swap** option, which is in Preview mode. This ensures that all app settings are swapped across the board. We can also configure slot-specific settings. The settings that are swapped are the general settings, app settings, connection strings, handler mappings, public certificates, and web job content. The list of what is not swapped is a little longer: publishing endpoints, custom domain names, non-public certificates, TLS/SSL, scale settings, web job schedulers, IP restrictions, Always On diagnostic settings, and **cross-origin resource sharing (CORS)**.

If we want to configure the deployment slots in Azure, we select **Deployment slots** under **Deployment**, and we can add more than one slot, as shown in the following figure:

Figure 7.28 – Deployment slots

The **Deployment slots** feature in App Service enables you to preview, manage, test, and deploy your different development environments.

To enable multiple deployment slots, the application must be running in the Standard, Premium, or Isolated tier. We select **Deployment slots** then **Add Slot**:

Figure 7.29 – Add Slot in Deployment slots

In the **Add a slot** dialog box, put a name for the slot and choose whether to clone the app configuration from another deployment slot. Select **Add** to continue. We will add a Dev environment:

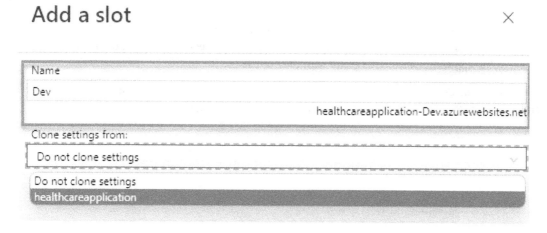

Figure 7.30 – Add a slot in Deployment slots

We can clone the configuration from an existing slot. Connection strings, web sockets, app settings, HTTP versions, language framework versions, and web sockets are the settings cloned.

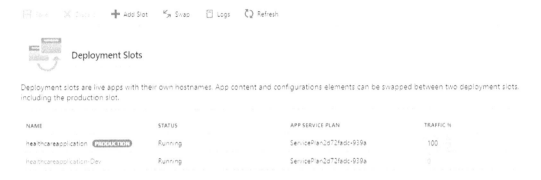

Figure 7.31 – Deployment Slots in App Service

A staging slot has an administration page like any other App Service app. You need to select the name and you will be redirected to the management page.

You can update the slot configuration. The application name is displayed in this format: `<app-name>/<slot-name>` and the application type is **App Service** (slot). A slot can also appear as a separate app within a resource group with the same label.

If we use Visual Studio 2022, we can deploy our application to this slot using the publishing tools and we can see the slot under the `Deployment Slots` directory, as shown in the following figure:

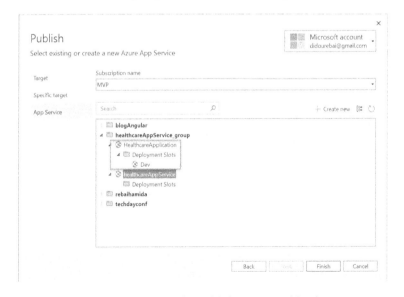

Figure 7.32 – Deployment slot publishing in Visual Studio 2022

Another feature of deployment slots is swapping operation steps. We will swap the **Dev** slot for the **Production** slot. The settings that are swapped are general settings, app settings, connection strings, hybrid connections, service endpoints, public certificates, web job content, and path mappings.

If you need more details about deploying to multiple staging slots, you can follow this documentation link: `https://learn.microsoft.com/en-us/azure/app-service/deploy-staging-slots`.

To swap **Dev** to **Production**, we will follow these steps:

1. On the **Deployment slots** page, select the **Swap** button.

2. Select the source, **Dev**, and the target slot, **Production**, and check **Config Changes**.

3. Select **Swap** to start the swapping operation:

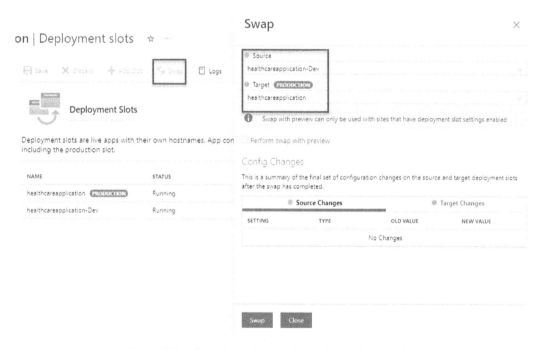

Figure 7.33 – Swap from the Dev slot to the Production slot

When the swap is completed, the content of the **Dev** slot will be on the **Production** slot.

In this section, we explored Azure App Service deployment slots and swapping between two slots.

Summary

In this chapter, we learned about the Azure App Service basics. We discussed Azure App Service and ASEs, and we created a new App Service instance. We also created and configured web app settings in Azure App Service, and to deploy the application on Azure App Service, we used Visual Studio 2022 and automated deployment. Then, we learned how to scale apps in Azure App Service, and finally, we explored Azure App Service deployment slots.

Azure App Service integrates with Docker in order to deploy containerized applications. In the next chapter, we will talk more about containers and the different services that can be used to deploy on Azure. The chapter covers architectural design and implementation approaches using .NET and Docker containers.

Further reading

For more details about App Service, you can head to the Microsoft documentation at `https://learn.microsoft.com/en-us/azure/app-service/` and the learning path at `https://learn.microsoft.com/en-us/training/modules/host-a-web-app-with-azure-app-service/`.

Part 3 – PaaS versus CaaS to Deploy Containers in Azure

In this part of the book, we will cover the development and deployment of a containerized application.

This part comprises the following chapters:

- *Chapter 8, Building a Containerized App Using Docker and Azure Container Registry*
- *Chapter 9, Understanding Container Orchestration*
- *Chapter 10, Setting Up a Kubernetes Cluster on AKS*

8

Building a Containerized App Using Docker and Azure Container Registry

Containerization is one of the biggest trends in software development today. Containers democratize cloud technology because containerized applications can be lifted and shifted to the cloud. You should evaluate whether containers are the right technology choice for every application in your workload. In this chapter, we will explore the different options for building containerized applications and integrating them with Microsoft Azure.

First, we will cover how to develop and deploy containerized ASP.NET Core applications with Docker and Azure Container Registry (ACR) We'll also cover architectural design and implementation approaches using .NET and Docker containers.

In this chapter, we're going to cover the following main topics:

- Describing the development process for Docker-based applications
- Exploring Azure Container Registry
- Exploring Azure Container Instances
- Exploring the elements of a Dockerfile
- Exercise 1 – deploying Docker containers on an Azure VM
- Exercise 2 – deploying Docker containers on Azure Container Registry

Describing the development process for Docker-based applications

Docker is a tool for building, deploying, and running applications using a containerization approach. These containers are lighter and take less time to launch compared to traditional servers. These containers also increase performance and reduce costs, while providing good resource management. Another advantage of using Docker is that you no longer need to pre-allocate memory to each container.

Docker was created to simplify the process of setting up environments where an application will run and is developed.

The development environment for Docker-based applications

Microsoft offers multiple tools for developing Docker applications:

- **Visual Studio (Windows and macOS)**: For Windows, if we want to create applications using .NET 6 or later, we can use Visual Studio 2022 version 17.0 or later. Different tools for Docker are already installed, and we can debug and run any application on a single or multiple containers locally in an installed Docker host. For Mac, this **integrated development environment** (**IDE**) is an evolution of Xamarin Studio on macOS. For a .NET 6 application, we need to install at least version 8.4 or later. It is a simple tool, with a user-friendly interface, and is powerful enough to create Docker-based applications.

- **Visual Studio Code (Windows, macOS, and Linux)**: We can use this IDE with a Docker **command-line interface** (**CLI**). It is a cross-platform editor because we can use it in Windows, macOS, or Linux (if you are using DEB files, you would need Debian or Ubuntu, and if you are using `.rpm` files, you would need Red Hat, Fedora, or SUSE). It includes Docker extensions such as IntelliSense, which allow us to create Dockerfiles, and we can use commands to debug and run Docker-based applications. To install Visual Studio Code, you can use this link: `https://code.visualstudio.com/download`.

We need to install Docker Desktop to be able to build applications for Windows and Linux with a single Docker CLI. We can install it on Windows using `https://docs.docker.com/desktop/install/windows-install/` or on macOS using `https://docs.docker.com/desktop/install/mac-install/`.

In this chapter, we will use a web application in .NET 6, but you can still use .NET Framework or any open source project if you want to develop a containerized .NET application. The target of the container can be Windows or Linux; it depends on .NET Framework.

Before describing the development process of building an application based on Docker, we need to understand the difference between a Docker container and a Docker image.

Docker containers versus Docker images

We will start by talking about containers, but before that, we need to look at virtualization and **virtual machines (VMs)**. VMs enable us to have multiple **operating systems (OSes)** in a single set of hardware. This has two main advantages:

- The first is **effective resource allocation**. When two VMs share the same hardware, each VM can use underutilized resources within the hardware.

- The second benefit of VMs is **isolation**. Applications running in separate VMs do not have access to each other's data.

The idea of virtualization is taken even further in the case of containers. When VMs virtualize hardware, containers virtualize the OS. Compared to VMs, containers are more portable and resource-efficient because process isolation in VMs is done at the hardware level, but in Docker containers, the process isolation is done on the OS.

VMs require more resources to build and run an application, which is not the case with Docker containers. If we want to optimize the creation of a VM, we need to use an automation engine, such as configuration management using Ansible, Chef, or Puppet, but it can be time-consuming, unlike the creation of Docker, which is simple and very quick to set up. Even customizing a VM is not easy compared to setting up a custom container.

Container images are typically smaller than VM images. A unit of isolation in VMs is a VM image. For containers, the equivalent unit is called a container image. Multiple containers can run on the same OS while still running as separate isolated processes. VM images are hosted in hypervisors, such as Hyper-V, KVM, or VMware. In turn, a Docker container can use just one machine, share its kernel, and virtualize the OS to run more isolated processes. So, Docker containers are lightweight.

Let's explore the Docker ecosystem to understand the different elements inside a Docker environment. The following figure presents the Docker architecture, which includes our **Docker Client**, **Docker Server**, and **Docker Registry**:

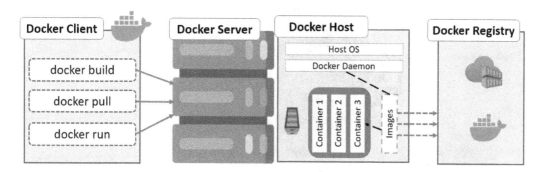

Figure 8.1 – Docker architecture

Let's look at each of these elements in detail:

- **Docker Client**: This communicates with the Docker daemon to execute specific tasks, such as running a container or building one or more images.

- **Docker Daemon**: This is responsible for building, running, and distributing Docker containers. We can run the Docker client and Docker daemon in the same system, or we can just connect the Docker client to a remote Docker daemon. The communication between the Docker client and the Docker daemon is ensured by a REST API. We can use **Docker Compose** as a Docker client, allowing us to work with applications using more than one container.

- **Docker Host**: This includes the Docker daemon and Docker objects.

- **Docker Server**: This is considered a compute instance and, like a VM, is a local machine that is responsible for running the daemon process. The daemon manages the different services sent by Docker and the different objects.

- **Docker objects** are containers, images, plugins, volumes, and networks. To manage the different Docker services, a daemon can communicate with more than one daemon.

- **Docker Registry:** This stores different Docker images. A registry is like a bookshelf where you can store and retrieve images to create containers that run services and web apps. We can store an image in a public registry, so we can use **Docker Hub**, which is maintained by Docker, or a private registry such as **Azure Container Registry**. Google and AWS offer several container registries, including **Amazon Elastic Container Registry** (**Amazon ECR**) and **Google Container Registry**, but you can also use private Docker registries on-premises and not just on a public cloud.

Containerized applications run on container hosts, which run on OSes (Linux or Windows). Therefore, containers have a much smaller footprint than VM images. As shown in the following diagram, each container can run any web application or service. In this example, the Docker host is a container host, and **App 1**, **App 2**, **Service 1**, and **Service 2** are containerized applications or services:

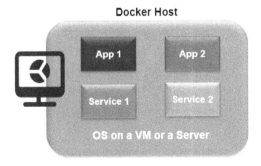

Figure 8.2 – Containerization in a Docker host

Container images are hosted in container engines. The most popular of them is Docker. Once you have an environment that has a container engine installed, you can run any container image on it, and it always behaves the same. This enables you to build your applications locally and ship the same container image, from staging to production.

A Docker image is like a snapshot in other types of VM environments. It is a record of a Docker container at a specific point in time. Docker images are also immutable. While they can't be changed, they can be duplicated, shared, or deleted. This feature is useful for testing new software or configurations because whatever happens, the image remains unchanged.

Containers require the existing runnable images to exist. They are dependent on images because they are used to construct runtime environments and are needed to run an application.

By default, if the image specified in the docker run command does not exist on the Docker host, the daemon will download this image from the public Docker registry – for example, from Docker Hub. If we need to create a new image, we can execute a set of command lines from a Dockerfile:

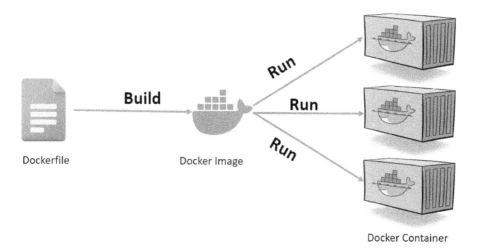

Figure 8.3 – Docker images and Docker containers

Now, we understand the difference between a Docker image and Docker containers and the elements that make up the architecture. In the next section, we'll have a look at the life cycle of a containerized application in Docker.

The containerized application life cycle in Docker

Before talking about the application life cycle in Docker, we will discuss the Docker container life cycle.

The life cycle of a container starts with creating a container from an image, at which point its status is **Created**. We can run this container if needed so that the status will be **Running**. If we want to delete a container, we can pause it. If we need to run the container again, we proceed by un-pausing. If we don't need the container, we can change its status to destroy it. The following diagram presents the different statuses of a Docker container:

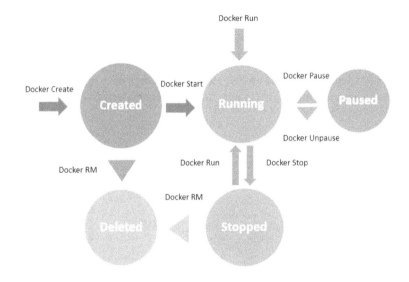

Figure 8.4 – Docker container life cycle

The following command can be used to stop a certain Docker container with the container ID:

```
$ docker stop CONTAINER_ID
```

The following command can be used to restart a certain Docker container with the container ID:

```
$ docker restart CONTAINER_ID
```

The journey of a containerized application life cycle begins by creating your application, and you can use any IDE (Visual Studio 2022 or Visual Studio Code). For this, we need to create a Dockerfile. A Dockerfile is like source code for Docker images. After we have written our Dockerfile, we must build it. A Dockerfile is built using the `docker build` command. The result of the build is a Docker image, which represents our application. It contains everything needed for our application to run. We can shift the container image anywhere where Docker is installed. Finally, we can run the application using `docker run`. We can see this process in the following figure:

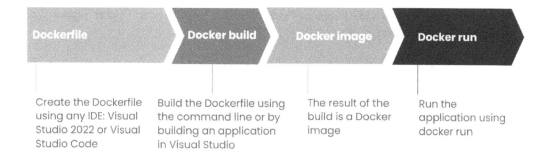

Figure 8.5 – Docker build flow

A Dockerfile is composed of a base image. There are dozens of official images available from Docker Hub, a repository of Docker images. For security issues, it's recommended to use one of the official base images so that you have the latest updates and security configurations taken care of. A Dockerfile is built in the context of the folder in the filesystem, where the Dockerfile is located.

You need a Dockerfile for each custom image you build. Whether you deploy automatically from Visual Studio or manually using the Docker CLI (the docker run and docker compose commands), you need a Dockerfile for each deployed container. For an application that includes a single custom service, you only need a single Dockerfile, but if it includes more than one service – for example, in a microservices architecture – you have to create a Dockerfile for each service.

The Dockerfile will be in the root folder of your application. We can add it by default when we create a new application in Visual Studio 2022, as presented in the following figure:

Additional information

ASP.NET Core Web App   C#   Linux   macOS   Windows   Cloud   Service   Web

Framework ⓘ

| .NET 6.0 (Long-term support) | ▾ |

Authentication type ⓘ

| None | ▾ |

☐ Configure for HTTPS ⓘ

☑ Enable Docker ⓘ

Docker OS ⓘ

| Linux | ▾ |
| Windows |
| Linux |

Figure 8.6 – Enabling Docker support using Visual Studio 2022 when creating a new application

Adding Docker support to an existing application is simple. In Visual Studio 2022, we must right-click on the application where we would like to add the Dockerfile. Next, we must select **Docker Support...**, as shown in the following figure:

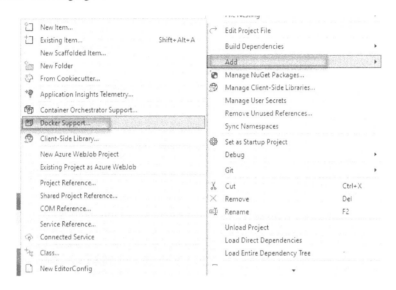

Figure 8.7 – Enabling Docker support in Visual Studio 2022 for an existing application

As shown in the following figure, select the target OS as Linux or Windows:

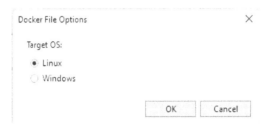

Figure 8.8 – Docker File Options in Visual Studio 2022

A new Dockerfile will be added to the solution.

> **Important note**
> Docker images are built automatically in Visual Studio 2022. You cannot explicitly create an image because it is created automatically when running the application by pressing *F5* (or *Ctrl* + *F5*).

If you are using **Visual Studio Code**, to generate a Dockerfile inside your application, you can use the Command Palette (*Ctrl + Shift + P*). Search for the **Docker: Add Docker Files to Workspace…** command:

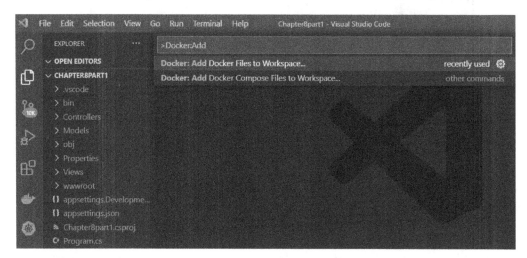

Figure 8.9 – Adding Docker files in Visual Studio Code

We have to select the application platform, as shown in the following figure:

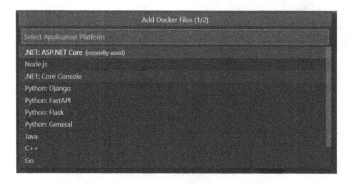

Figure 8.10 – Creating a Dockerfile in Visual Studio Code – Select Application Platform

Next, we must select the OS for our container (Linux or Windows):

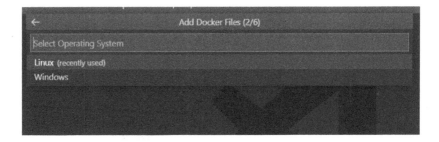

Figure 8.11 – Creating a Dockerfile in Visual Studio Code – Select Operating System

We will be requested to select or update the ports where the application will listen; in the end, we will have the option to create a **Docker Compose** file. This is used to define and run multi-container Docker applications. The Dockerfile, along with another file type called .dockerignore, are added to the workspace. The Dockerfile will include the different basic instructions that allow you to run your application on a container. The file will be added to the root folder of your application, as presented in the following figure:

```
FROM mcr.microsoft.com/dotnet/aspnet:6.0 AS base
WORKDIR /app
EXPOSE 5026

ENV ASPNETCORE_URLS=http://+:5026

FROM mcr.microsoft.com/dotnet/sdk:6.0 AS build
WORKDIR /src
COPY ["Chapter8part1.csproj", "./"]
RUN dotnet restore "Chapter8part1.csproj"
COPY . .
WORKDIR "/src/."
RUN dotnet build "Chapter8part1.csproj" -c Release -o /app/build

FROM build AS publish
RUN dotnet publish "Chapter8part1.csproj" -c Release -o /app/publish /p

FROM base AS final
WORKDIR /app
COPY --from=publish /app/publish .
ENTRYPOINT ["dotnet", "Chapter8part1.dll"]
```

Figure 8.12 – The Dockerfile structure

Once the Dockerfile has been created, we must configure it, open it, and use an ENV instruction to add an environment variable to our service container image. We will use port 5056 for the application, so we will set the value of the ASPNETCORE_URLS variable to listen on port 8.0. We can build it into a Docker image. To do that, we can use the Command Palette (*Ctrl + Shift + P*) and search for Docker Images: Build Image.

Now that our application is built into a Docker image, we can run it anywhere with the Docker client installed. We will use Visual Studio Code to run it. Select the Command Palette (*Ctrl + Shift + P*) and type Docker run, as shown in the following figure. We can use different methods: an interactive method or use the Azure CLI:

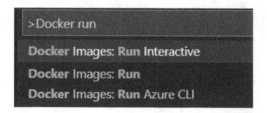

Figure 8.13 – Running Docker images using Visual Studio Code

Running with the interactive method

With this method, users will run a container from an existing Docker image and manually make any specific changes to the environment before saving the image. This method is the simplest and easiest for creating Docker images. When we use Visual Studio Code, we will be requested to select an existing image.

Running using the Azure CLI

We can only run the Azure CLI when running Linux-based containers.

Running using a Dockerfile

This approach requires us to make a plain text Dockerfile similar to the batch script. This process is more difficult and time-consuming, but it does well in continuous delivery environments. The method includes creating the Dockerfile and adding the commands needed for the image. Once the Dockerfile has been started, the user sets up a .dockerignore file to exclude any files not needed for the final build. The .dockerignore file is in the root directory.

In our example, we will run using the interactive method, and you can use localhost to open your web application. By default, Docker assigns a random host port to the port provided by the container (the container port). In our case, the exposed port is 5026 and the link is http://localhost:5026/.

If we need to debug the application, we can add a breakpoint in the index method – inside HomeController, for example – and start debugging using *F5*. Alternatively, at the top of Visual Studio Code, we can select **Run** after **Start Debugging**. The configuration should be similar to what's shown in the following figure, where we have selected **Docket .NET Core Launch**:

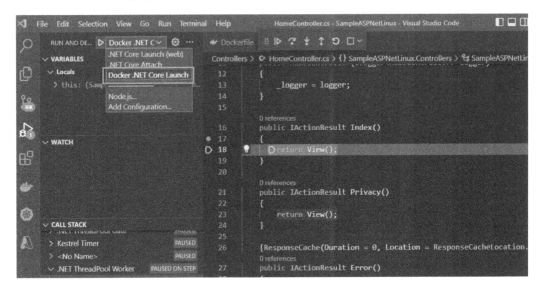

Figure 8.14 – Debugging an application

We can use the Docker command line in the Terminal instead of using Visual Studio 2022 or Visual Studio Code.

We are going to launch Docker and open a terminal session or PowerShell terminal. To do this, we must run the docker images command to list all the available Docker images in our environment:

```
$ docker images
```

We can see that the different images are created with a tag, image ID, size, and date:

Figure 8.15 – Docker images in the local environment

If we want to build a Docker image, we can use this command line:

```
$ docker build -t yourimage .
```

To run the Docker container images, we can use image_name:tag_name with the docker run command. This starts a shell session with the container that was launched from the image. We're going to use the docker run command:

```
$ docker run -it --rm -p 8000:80 --name theapplicationname
yourimage
```

If the tag name is omitted, Docker uses the most recent version of the image. The image should appear listed in the results. We can open Docker Desktop or use the docker images command line in any terminal session, such as PowerShell:

Figure 8.16 – Displaying Docker images in Docker Desktop

If your application is using a microservices architecture, it will include multiple services. For each service, we must create an image. Integrating microservices with the applications running on containers is easy. Each tier of a multi-tier application running on Docker behaves as an independent container and can be used to integrate microservices with the application.

We need to create Docker images, deploy containers locally (using the local Docker host), and run, test, and debug all containers before using any cloud service.

Docker Compose

If we have multiple containers, we might want to build all the images together using a single command line. In this case, we will use the docker-compose up –build command line, and we will use the metadata exposed in the associated docker-compose.yml files.

A Docker Compose file is used to define and run multi-container Docker applications. We can create a YAML file to configure the application services (`docker-compose.yml`), which will also configure the dependency relations of the different services and their runtime configuration.

> **Important note**
> **Yet Another Markup Language** (**YAML**) is a data serialization language commonly used to create configuration files.

To create or start all services with a single command, we can use `docker compose up`, but if we need to stop all services, we can use the `docker compose down` command. Using these, we can scale up selected services when needed.

We can use the `docker-compose run` command to run the one-off or ad hoc tasks that are required to be run, as per application needs and requirements. We need to provide a service name that we would want to run; based on that, the command line will only start the containers for the services that the running service depends on. Using the `run` command, we can run our tests or perform any administrative tasks, such as removing/adding data from/to the data volume container.

We can use the `docker-compose start` command to restart the containers that were created and stopped. This command doesn't create any new Docker containers by itself.

Containerizing a sample ASP.Net Core 6 and SQL Server with Docker Compose

To understand more about the use of `docker-compose`, we will use a sample ASP.NET Core application that includes an API to manage the different operations for booking a medical appointment, a web application that will be the user interface for patients, and another web application that will be used by doctors as a dashboard to display, delete, or confirm appointments.

The API uses Entity Framework 6 as an object-relational mapper to communicate with a database, which allows us to query and manipulate data from the database using objects. The database will be running as a container, which is named `sqldb`. We will define it in the `docker-compose.yml` file. This container runs a SQL Server instance.

To run the container separately instead of using `docker-compose`, we will use this command line:

```
docker run -e 'ACCEPT_EULA=Y' -e 'SA_PASSWORD=My@PassWord \
-p 5433:1433 --name sqldb --hostname sqldb -d \ mcr.microsoft.
com/mssql/server:2019-latest
```

- `-e "ACCEPT_EULA=Y"`: We use Y to say yes, which means that we confirm acceptance of the end user licensing agreement.
- `--name sqldb`: This is the name of the container. It is optional because, without this parameter, a random name will be generated.

- -d: This is used to run the container as a daemon in the background.

- `mcr.microsoft.com/mssql/server:2019-latest`: This is the SQL Server Linux container image.

The following figure presents the different interactions between the API, the database, and the web applications:

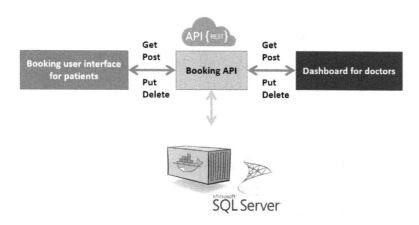

Figure 8.17 – Application and API interactions

The problem is that if we run every application in a container, they can't communicate – for example, if we run the API and the database and we start sending an HTTP request to the API, do we get any response? No – if we don't use `docker-compose` or any orchestrator, we are not able to get data and send a response.

We can write `docker-file.yml` by adding a YAML file to the root of the solution, but we can generate it and update the database element using Visual Studio Code or Visual Studio 2022.

If we use Visual Studio 2022, we can generate a `.yaml` file to orchestrate containers using container orchestrator support. Right-click on the project, then select **Add | Container Orchestrator Support…**:

Application Insights Telemetry...	Build Dependencies ▶
Container Orchestrator Support...	Add ▶
Docker Support...	Manage NuGet Packages...
Client-Side Library...	Manage Client-Side Libraries...
New Azure WebJob Project	Set as Startup Project
Existing Project as Azure WebJob	Debug ▶
Reference...	Git ▶
Service Reference...	Cut Ctrl+X

Figure 8.18 – Adding container orchestrator support in Visual Studio 2022

We will do that with all the projects inside the .NET solution.

A new node in the **Solution** explorer will be added and will contain the docker-compose.yml files, as shown in the following figure:

Figure 8.19 – The docker-compose.yml files in Visual Studio 2022

This is an example of a docker-compose.yml file. We added the block related to the database container at the end, and we also added depends_on, including the name of the database container in the API block that will use the database:

```
version: '3.4'
services:
  healthcaresolution.dashboard:

    image: ${DOCKER_REGISTRY-}healthcaresolutiondashboard
    build:
      context: .
      dockerfile: HealthcareSolution.Dashboard/Dockerfile
    ports:
            - "8000:80"
    depends_on:
            - sqldb


  healthcaresolution.bookingapi:

    image: ${DOCKER_REGISTRY-}healthcaresolutionbookingapi
    build:
```

```
    context: .\HealthcareSolution.BookingAPI
    dockerfile: Dockerfile
  ports:
        - "8000:80"
  depends_on:
        - sqldb


healthcaresolution.bookingui:

  image: ${DOCKER_REGISTRY-}healthcaresolutionbookingui
  build:
    context: .\HealthcareSolution.BookingUI
    dockerfile: Dockerfile
  ports:
        - "8000:80"
  depends_on:
        - sqldb


sqldb:
    image: mcr.microsoft.com/mssql/server:2019-latest
    environment:
    - SA_PASSWORD=My@PassWord
    - ACCEPT_EULA=Y
    ports:
    - "5433:1433"
```

We added a name for every part. This file includes static configuration data for every container, and we have the name of the container, the image, the environment, and the ports.

To run the docker-compose.yml file, we can just click on **Docker Compose** in Visual Studio 2022:

Figure 8.20 – Running a docker-compose.yml file in Visual Studio 2022

But if you prefer using the Docker Compose command line, you can open any Terminal session and use docker-compose up, as shown in the following figure, to deploy a multi-container application:

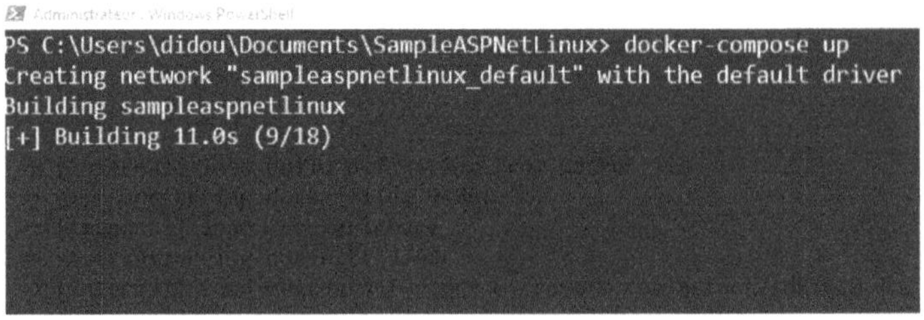

Figure 8.21 – The results of the docker-compose up command

In this section, we presented the Docker app life cycle. We started by creating the application or using any application that will be pushed onto a container. We added Docker support using Visual Studio 2022 and Visual Studio Code. Then, a Dockerfile was generated, which we ran in our containers; for multiple containers, we used the docker-compose.yml file.

We used Docker Desktop to display the local images and containers, but if we need to deploy our container in Azure, we can use **ACR**, which is the topic of the next section.

Exploring Azure Container Registry

ACR is an Azure service that is used for creating private Docker registries.

It is similar to Docker Hub but offers a few unique benefits: ACR runs in Azure, is highly scalable, and provides enhanced throughput for Docker pulls that can span many nodes concurrently.

ACR facilitates version control for container images. Developers push container images to container registries, and administrators pull those images from registries. A container registry provides image versioning and helps you to manage tagged versions. ACR is a fundamental building block that enables the distribution of container images. Instead of uploading container images directly from your development environment to your server, continuous deployment tools are often instrumented to pull images from container registries and put them on servers. ACR has many features for security and high availability. First, access to container registries can be controlled using **Azure Active Directory** (**Azure AD**) and firewalls. Then, repositories can be replicated across Azure data center regions. Also, you can sign your image with Docker Content Trust.

Instead of building container images on developer machines, you can use ACR to run docker build commands in the cloud. ACR then inherits the Dockerfile and builds context. No input is required. You can build it into a container image and, finally, publish it to the registry.

ACR also supports automatically triggered tasks. Tasks can be scheduled or run based on external metrics, such as source code modifications. For example, you can configure ACR to rebuild your application image when an update to the base image used becomes available on Docker Hub.

Exploring Azure Container Instances

Azure Container Instances (ACI) is a fully managed container hosting service in Azure. While ACI provides some of the basic capabilities of container orchestrators, it should not be considered a lightweight container orchestrator. Rather, ACI can be used to complement the use of container orchestrators in the cloud. For example, virtual notes in Azure Kubernetes Service use ACI for fast, on-demand capacity. ACI comes with a full set of features and interfaces native to Azure capabilities. Applications hosted in ACI cannot be deployed to Azure Virtual Networks and integrated with on-premises networks using a side-to-side VPN.

ACI is more flexible in the deployment footprint than deploying applications to VMs directly. While Azure VMs require you to pick from a list of predetermined sizes, ACI lets you freely choose the CPU count and memory allocation for your application. ACI even supports direct integration with the SMB file share service of Azure storage. Lastly, ACI is simply an incredibly fast way to start running containers in Azure. You can get your container application running in a matter of seconds, with a single Azure CLI command.

To create a new Azure Container Instance, we can use the Azure portal, the Azure CLI, Azure PowerShell, or the Azure ARM template. Let's look at the steps:

1. We will create a new Container Instance using the Azure portal. Select **Create a Resource | Container category**, then click **create Container instances**.

2. Under **Project details**, we must select the subscription and the resource group, while under **Container details**, we must introduce the container name, the region, and the availability zone, as well as select an image source. For the image source, you can choose from **Quickstart images**, **Azure Container Registry**, or **Other registry**, such as Docker Hub or an on-premises registry:

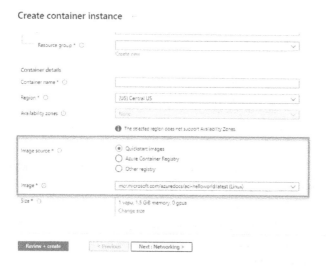

Figure 8.22 – Creating a container instance in the Azure portal

There are more tabs that we can update according to our needs. Click **Review + create** to create our container instance.

Exploring the elements of a Dockerfile

A Dockerfile contains all the instructions needed to build a Docker image. This includes copying our source, making changes to the base image such as adding new packages or updates, exposing necessary reports, and configuring environment variables.

The Add, CMD, Entry point, ENV, EXPOSE, FROM, MAINTAINER, RUN, USER, VOLUME, and WORKDIR commands are available in a Dockerfile.

This is an example of a Dockerfile:

```
FROM mcr.microsoft.com/dotnet/aspnet:6.0 AS base
WORKDIR /app
EXPOSE 80

FROM mcr.microsoft.com/dotnet/sdk:6.0 AS build
WORKDIR /src
COPY ["HealthcareSolution.Dashboard/HealthcareSolution.Dashboard.csproj", "HealthcareSolution.Dashboard/"]
RUN dotnet restore "HealthcareSolution.Dashboard/HealthcareSolution.Dashboard.csproj"
COPY . .
WORKDIR "/src/HealthcareSolution.Dashboard"
RUN dotnet build "HealthcareSolution.Dashboard.csproj" -c Release -o /app/build

FROM build AS publish
RUN dotnet publish "HealthcareSolution.Dashboard.csproj" -c Release -o /app/publish /p:UseAppHost=false

FROM base AS final
WORKDIR /app
COPY --from=publish /app/publish .
ENTRYPOINT ["dotnet", "HealthcareSolution.Dashboard.dll"]
```

Figure 8.23 – A Dockerfile sample

Let's understand these elements of a Dockerfile in detail:

- FROM mcr.microsoft.com/dotnet/sdk:6.0: The Dockerfile always starts with the FROM command. It initializes a new build stage and sets the base image that we are going to build upon.

- WORKDIR /source: The WORKDIR command simply sets the current working directory inside our image. In this case, it is the /source folder, which is the root of our solution.

- COPY *.sln . and COPY aspnetapp/*.csproj ./aspnetapp/: The COPY command is used to copy all the files from the local system to the current working directory of the image. In this example, we will use a .dockerignore file that the COPY command will look up when it starts copying the files.

- `RUN dotnet restore`: The RUN command runs any command in a new layer and commits it to the base image. In our case, we will restore the packages for our solution, similar to running it locally, but it will run inside the image.

- `RUN dotnet publish -c release -o /app --no-restore`: We will switch the current directory to `/publish`.

- `ENTRYPOINT ["dotnet", "aspnetapp.dll"]`: The ENTRYPOINT command allows us to configure the container to run as an executable.

Because we will copy all files to the Docker image on every build, we will use the `.dockerignore` file. We can also select which files and folders we don't want to copy every time.

Now that our image is ready, we will build it and store it using ACR.

Exercise 1 – deploying Docker containers on an Azure VM

After preparing the environment in a VM (Windows or Linux), we can publish our application to this machine using Visual Studio 2022.

Right-click on the project in Visual Studio. Then, select **Publish** and then **Azure**. We will see **Azure Virtual Machine** in the list of Azure services, where we can host our application:

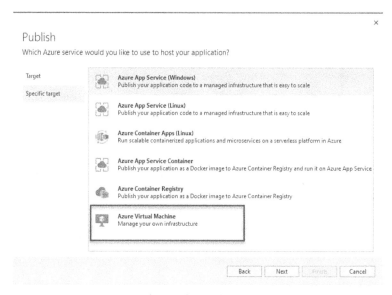

Figure 8.24 – Deploying the application to an Azure VM

We need to select the instance after the application has been deployed if the VM includes Docker.

Exercise 2 – deploying Docker containers on Azure Container Registry

We will start with the first step, which is creating a Docker image of a web application we will create from an existing Visual Studio template. Alternatively, we can use Microsoft ASP.NET Core images to push and run our image locally. You can find the .NET samples here: `https://hub.docker. com/_/microsoft-dotnet-samples/`.

To create an ACR, we can use multiple methods: the Azure portal, the Azure CLI, Azure PowerShell, and the Azure ARM template.

Creating an Azure Container Registry with the Azure portal

Let's look at the steps:

1. To get started, we will access the Azure portal. Click on **Create a resource**:

Figure 8.25 – Creating a resource in the Azure portal

2. Select **Containers** to the left of the window. Next, choose **Container Registry**:

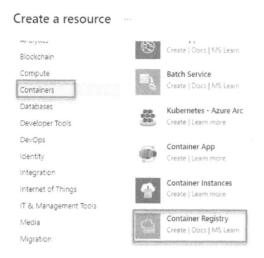

Figure 8.26 – Containers | Container Registry

3.   Click on **Create** to create a new container registry.

4.   On the next screen, you need to configure your registry. Select the subscription, the name of the registry, the location, and the SKU – for this example, I chose the basic SKU. Check **Enabled** for **Availability zones**, and then click on **Review + create**. Select **Create** when the validations have been successfully passed:

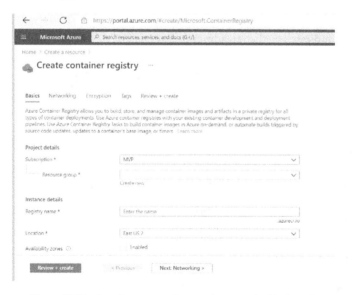

Figure 8.27 – Creating a container registry on the Basics tab

We have more than one tab, including **Networking**, **Encryption**, and **Tags**.

If you select the basic or standard SKU, in the **Networking** tab, you will not be able to configure **Connectivity** to connect to this registry either publicly, via public IP addresses, or privately, using a private endpoint. Only a premium SKU allows this.

On the **Encryption** tab, ACR service encryption protects data by encrypting images and other artifacts when they're pushed to your registry and automatically decrypts when you pull them. However, **Customer-Managed Key** is only available for a Premium SKU.

Creating an ACR with the Azure CLI

Let's look at the steps:

1. If you don't have a resource group, you need to create a resource group by using the az group create command:

    ```
    az group create --name <resource-groupe-name> --location
    <location>
    ```

2. Next, you must create an ACR instance in your resource group by using the az acr create command:

    ```
    az acr create --resource-group <resource_group_name>
    --name <registry_name> --sku Basic --admin-enabled true
    ```

If we want to log in to the container registry, we have to be administrators, so we need to add -admin-enabled true to indicate whether an admin user is enabled.

Pushing your image to ACR using the Docker CLI

Even though we are administrators on the container registry, we will log in to the ACR we created earlier by using the az acr login command:

```
az acr login --name <registry_name>
```

This command returns a **Login Succeeded** message once completed.

Now, to use our application container image with ACR, we have to tag the image with the login server address of our registry. To do that, follow these steps:

1. Use the docker images command to view your list of local images:

    ```
    $ docker images
    ```

2. Get the login server address for the ACR by using the `az acr list` command:

```
az acr list --resource-group <resource_group_name>
--query "[].{acrLoginServer:loginServer}" --output table
```

3. Now, you need to tag the application image with the login server address of your registry from the previous step. This will add an alias of the application image with a complete path to your registry:

```
docker tag yourtagtoadd<registry_login_server>/
yourtagtoadd:01
```

4. You can use `docker images` to verify your tag.

5. Use the `docker push` command to push the application image to your container registry:

```
docker push <registry_login_server>/yourtagtoadd:01
```

6. Finally, you need to validate whether the image has been uploaded to your registry using the following command line:

```
az acr repository list --name <registry_login_server>
--output table
```

You can use the interactive Azure publishing tool in Visual Studio 2022 or Visual Studio Code.

Deploying the application to a container registry using Visual Studio 2022

Right-click on the project and select **Publish**. Then, select **Azure | Azure Container Registry**:

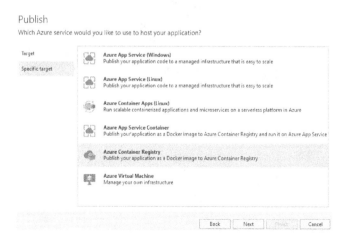

Figure 8.28 – Publishing to ACR using Visual Studio 2022

We will select **Azure Container Registry**, which has already been created in Azure. We can also create it using Visual Studio. Finally, we will publish the application to ACR.

Deploying the application to a container registry using Visual Studio Code

Follow these steps to deploy our application to ACR using Visual Studio Code:

1. Open **Docker Explorer** on the left with the Docker icon, and then select **Connect Registry** under the **Registries** group. We need to follow the prompt.

2. Select the **Azure** provider and provide the credentials to connect to the registry. If you get an error, check whether you have already installed the Azure Account extension.

3. We will see the registry under **REGISTRIES**:

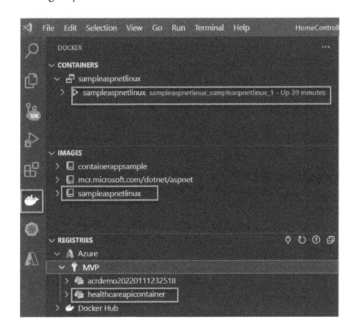

Figure 8.29 – Deploying an application to ACR

4. We can tag the image, though this is optional, and we will push the image, as shown in the following figure. Right-click and then select **Push...**:

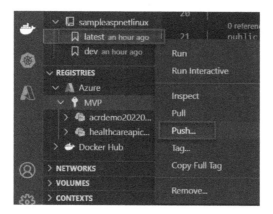

Figure 8.30 – Pushing the image with the tag to Azure Container Registry

You can pull the image, delete it, and also run it using Visual Studio Code.

Summary

In this chapter, we learned about the development process for Docker-based applications, and the fundamentals and terminologies of Docker and ACR. We learned about ACI and discussed how it differs from ACR. We understood the container design principles and how we can manage states and data in Docker applications. We also built and managed containers with tasks.

We learned that a Dockerfile is a text file that includes the different instructions presented as a command to build images automatically, and we explored the different elements of a Dockerfile for use after deployment. Finally, we deployed Docker containers using different environments, first with an Azure VM on Windows and Linux and second on ACR.

If you have multiple containers, is it easy to manage them? Do we need another tool to orchestrate them? This is the topic of the next chapter. We will learn about container orchestration when building microservices and multi-container applications, ensuring that we get the benefits of high scalability and availability.

Further reading

You can find more information about Docker containers in Microsoft eBooks:

- *.NET Microservices: Architecture for Containerized .NET Applications*:

  `https://docs.microsoft.com/dotnet/architecture/microservices/`

- *Modernize existing .NET applications with Azure cloud and Windows Containers*:

  `https://docs.microsoft.com/dotnet/architecture/modernize-with-azure-containers/`

9
Understanding Container Orchestration

A container is a packaging mechanism that solves multiple problems related to installation environment issues. When an application is traditionally deployed from a development environment to a production environment, the application may not work despite it having worked on the developer's machine. We realize that we have a problem and the application is broken. The containerization approach resolves this problem by creating a strict boundary between the infrastructure and software stacks used by applications. We don't need to add external dependencies to the container – only internal dependencies such as the framework and runtimes are added.

Despite containers being ephemeral and lightweight, they do sometimes require a massive effort to run in production, especially if we use microservices deployed in multiple containers and need to communicate between them. If we built a large-scale application, we would have to face the complexity of managing many different containers manually, and this is where the orchestrator comes to the rescue.

In this chapter, you will understand the importance of container orchestration when building microservices and multi-container applications to ensure access to the full benefits of high scalability and availability.

In this chapter, we're going to cover the following main topics:

- Container orchestration versus Docker
- Exploring Azure Kubernetes Service and Azure Container Apps
- Orchestrating microservices and multi-container applications

Container orchestration versus Docker

For a self-contained development environment, it's pretty easy to use a tool like Docker to launch a container and start working. This approach works to a limited extent on production systems, but it doesn't take full advantage of containers.

In fact, if you're containerizing an existing complex stack, separating functions into containers and keeping them on the same machine might be a good transition state. A fully containerized application stack is much easier to split into orchestration systems piece by piece. The basic job of a container orchestrator is to manage the state of containers to respond to incoming workloads. This includes many things such as load balancing and scaling, networking, storage, scheduling, deployment, and more.

We use orchestrators in a production-ready application, especially if the application is based on microservices and a complex distributed system deployed across more than one container. In a microservices approach, each microservice is independent in its model data so it is autonomous from development to deployment. In the **Software-Oriented Architecture** (**SOA**) approach, we can have more than one container because a traditional application can include more than one layer, but we will deploy the application in a single business package similar to a distributed system. In these systems, scaling out will be complex to manage, but an orchestrator can help us to manage a multi-container application.

Container orchestration makes it easier for development and operations (or DevOps) to manage this operational complexity by providing a declarative way to automate a lot of the work. This works well for DevOps teams and cultures that typically aim to be much faster and more agile than traditional software teams.

An orchestration platform can handle all of this dynamically. Autoscaling and load balancing are just parts of orchestration. Orchestrators often handle storage, scheduling, networking, and more. A configured orchestration platform provides a layer of abstraction so that the infrastructure almost fades into the background, and developers are free to focus more on code development.

In the previous chapter, we learned about the components of Docker. If we want to scale out our application across multiple Docker hosts via a distributed solution using multiple microservices applications, it is important to be able to manage all these hosts as a single cluster while being able to ignore the complexity of the underlying platform.

Container orchestration is a broad term for automating the life cycle of any container. Docker also includes **Docker Swarm**, the platform's native container orchestration tool that can autostart Docker containers.

Docker Swarm – a definition

Docker Swarm is the orchestration solution provided by Docker. A Docker swarm is a set of physical or virtual machines running Docker applications that are configured to form a cluster. Its orchestrator functionalities allow users to manage multiple containers deployed on multiple host machines. Once a group of machines is clustered, users can still run regular Docker commands, but they will be run by the machines in the cluster.

Kubernetes – a definition

Kubernetes is an orchestration solution that offers a different approach to that taken by Docker Swarm, targeted instead at top-down centralized infrastructure. Kubernetes clusters run containers built on Docker and other platforms.

The control plane in a Kubernetes cluster manages the workload and communication across the cluster. Containers run individual instances of a container engine on the same operating system kernel. The basic elements of a Kubernetes cluster are as follows. The most modern container engine uses the **Open Container Initiative (OCI)** container image format.

Containers are run on physical or virtual machines. A **Pod** is a logical unit of a Kubernetes workload. Containers that are located in the same Pod run on the same **node**, and they interact with each other using localhost rather than the network. Multiple pods of the same set of containers are grouped into a **ReplicaSet**. Replicas are abstracted into services to balance the workload across these replicas. This is similar to using a load balancer to balance traffic across multiple VMs.

Kubernetes cluster architecture

Let's take a look at the Kubernetes cluster architecture. We have two principal components, the **master node** and the **worker node**:

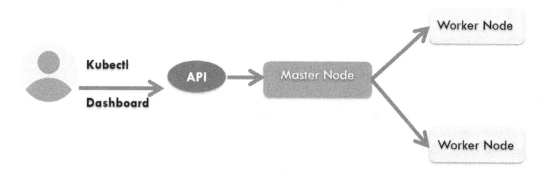

Figure 9.1 – Kubernetes cluster architecture

The **master node** is responsible for the overall management of a Kubernetes cluster. The master node coordinates all the activities in the cluster and communicates with the nodes in order to keep Kubernetes and your applications running.

The user interacts with the master node using the **kubectl** application, which is a command-line interface for Kubernetes. But the interaction with the API server can be done also by using the user interface (dashboard).

kubectl has a configuration file called `kubeconfig`. This file includes the server information and the authentication information to allow access to the API server.

Worker nodes are the nodes on which your applications run. The worker nodes communicate with the master node.

Let's examine the different interactions between the master and worker nodes in more detail:

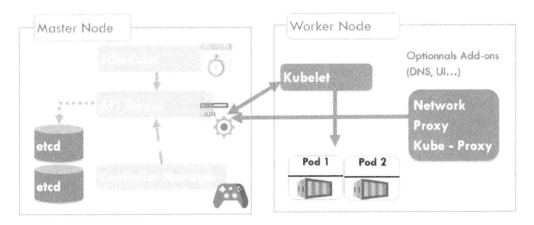

Figure 9.2 – Master node and worker node components

Worker nodes are agents that communicate with the API server to see whether Pods have been designated to nodes. They execute Pod containers via the container engine. They mount and run Pod volumes and secrets. And finally, they monitor the node states in a Pod and report back to the master node.

The master node comprises three components: the **Scheduler** , the API server, and the controller manager. The API server allows you to interact with Kubernetes APIs and is considered the frontend of the Kubernetes control panel. The **Scheduler** monitors Pods that have been created but do not yet have a node designation, and designates the Pod to run on a specific node. The **controller manager** runs controllers in a background thread running tasks in a cluster. Controllers can have multiple roles, including being responsible for the worker states, while replication controllers maintain the correct number of Pods for the replicated controller, and endpoint controllers join the services with the Pods together. The controller is compiled into a single binary.

The communication with worker nodes is handled by the **kubelet** process.

Worker nodes can have more than one Pod designated to them. By definition, a Pod is the smallest unit that can be scheduled for deployment in Kubernetes. This group of containers will share storage, Linux namespaces, IP addresses, and so on. They also share resources that are localized and always scheduled together. Once a Pod is deployed and running, the kubelet process communicates with the Pod to check its status and health, and **kube-proxy** forwards packets to the Pod from other resources that may communicate with the Pod.

Note that the worker nodes can be exposed to an external network such as the internet via the load balancer, and incoming traffic to the nodes is handled by kube-proxy.

When applications run on nodes, a basic structure is required to build a Kubernetes app. In the Kubernetes model, a Pod is the simplest unit you can interact with. We are able to create, deploy, and delete Pods, and each Pod represents one process running in your cluster. A Pod contains Docker application containers, storage resources, unique network IPs, and options that determine how containers run.

A Pod is a group of one or more containers; it is the smallest deployment unit in Kubernetes that represents a single instance of a containerized application that is tightly coupled and shares resources. Pods are designed as short-lived, disposable units.

A Pod can be used as a template in order to configure other deployment objects such as **ReplicaSets**, **Deployments**, and **StatefulSets**.

A **StatefulSet** is the workload API object used to manage stateful applications.

A **Deployment** provides declarative updates for Pods and ReplicaSets. Pods are created manually using a Pod manifest file or automatically by the controller when creating ReplicaSets, Deployments, and StatefulSets.

In the next section, we'll explore the most popular platform and software solutions for orchestrators: the Azure public cloud.

Exploring Azure Kubernetes Service and Azure Container Apps

Azure offers orchestration services to manage multiple containers. In the preceding chapter, *Chapter 8, Building a Containerized App Using Docker and Azure Container Registry*, we explored the use of Azure Container Registry to store and manage private container images. If we use a single container, we don't need any orchestration solution – we can simply use Azure Container Instances or Azure App Service as a container to run isolated containers. For multiple containers, we can use Azure Kubernetes Service and Azure Container Apps. In the following figure, we will present the different Azure services for containers:

Store

· Azure Container Registry

For a single container

· Azure Container Instance
· Azure App Service as a Container

For multiple containers

· Azure Kubernetes Service
· Azure Container App

Figure 9.3 – Container-based orchestrators in Azure

A Kubernetes cluster is a set of Docker hosts; the cluster deploys these hosts in a single virtual Docker host. A Kubernetes cluster allows you to deploy multiple containers in the cluster. This **Deployment** scales to any number of container instances. The cluster can handle a set of complex management tasks including scalability and health management.

Azure Kubernetes Service

The Kubernetes community is a growing community with the goal of sharing skills and expertise in deploying and managing container-based applications on Kubernetes.

This managed Kubernetes service allows you to deploy production-ready Kubernetes clusters in Azure environments.

There are open source container orchestration frameworks such as Kubernetes – for example, Docker Swarm, Red Hat OpenShift, Docker Compose, and HashiCorp Nomad.

Docker Swarm is a container orchestration tool native to the Docker platform and is used to manage containerized applications.

Docker Compose is a Docker orchestration tool for running multi-container applications on Docker using a compose file.

Red Hat OpenShift is a secure and enterprise-grade cloud-based container orchestration system based on Kubernetes at the backend. Red Hat OpenShift can be used as a platform-as-a-service and a Red Hat container orchestration engine.

HashiCorp Nomad is a cluster manager and **Scheduler** to deploy a containerized or legacy application across an infrastructure and a flexible workload orchestrator.

Azure Kubernetes Service (**AKS**) uses a set of open source orchestration and scheduling tools and provides a simplified way to create, configure, and manage clusters of virtual machines in Azure. The clusters are preconfigured to run containerized applications.

AKS optimizes your Docker cluster configuration. AKS allows you to choose the size and number of hosts and take advantage of various orchestration tools. Everything is handled by AKS.

In AKS, pools are groups of nodes with identical configurations. Cross-schema nodes are individual virtual machines running containerized applications. We use a deployment file (YAML file). This file includes several pods. A deployment has one or more identical Pods managed by Kubernetes. The manifest is the YAML file describing a deployment.

A **Deployment** provides declarative updates for Pods and ReplicaSets and bundles them (Pods and ReplicaSets) into a package that is capable of deploying your applications. The Kubernetes **Scheduler** ensures the availability of Pods or nodes to avoid any problems encountered from making the app completely unavailable as there are additional Pods scheduled on healthy nodes. A **pool** can contain several nodes. The following figure illustrates the AKS components:

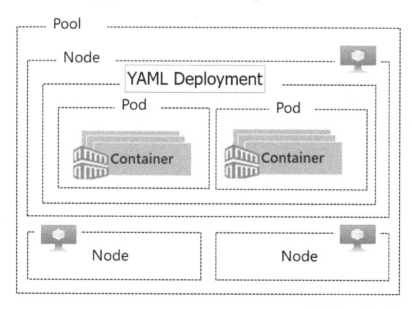

Figure 9.4 – Azure Kubernetes Service components

Let's take a look at a Kubernetes cluster where the master node controls the coordination of the cluster. Containers are deployed to the other nodes and managed as a single pool. This setup can scale to thousands of containers:

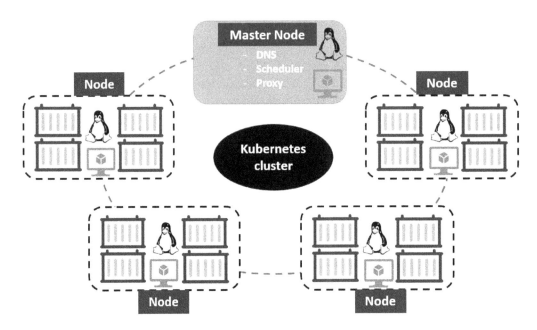

Figure 9.5 – Kubernetes cluster structure

In the Kubernetes cluster architecture, we have two basic components:

- The control plane provides the orchestration of the application workloads and the core Kubernetes services and nodes used to run the application workloads.

- Worker nodes can be physical machines or VMs. A node hosts Pods, which run one or more containers.

Kubernetes is a sophisticated, somewhat complex infrastructure to manage, update, observe, and maintain containers. There are platform-as-a-service offerings that handle some of these responsibilities, although that still leaves many tasks to be done. Azure Container Apps is another service that we can use to manage multiple-container setups.

Azure Container Apps

Azure Container Apps is a **serverless** container platform for building modern apps and microservices. It sits on top of an AKS instance.

With Azure Container Apps, we can deploy multiple containers without needing to deal with the underlying infrastructure. In fact, Azure Container Apps doesn't even expose Kubernetes APIs to its users.

If we deploy or update the containers in an Azure Container Apps instance, the service allows us to automatically create a snapshot of the application (called a revision). All these containers share the same application life cycle and resources, including compute, storage, and network, similar to Kubernetes. Different containers can communicate with each other. We have the integration of **Kubernetes-based Event-Driver Autoscaling** (**KEDA**) that allows us to scale (only horizontally as vertical scaling is not supported) the number of Pods related to a revision according to defined metrics, such as memory usage or the number of HTTP concurrent connections.

If the application doesn't receive any requests, the service will scale down by reducing the number of active Pods to 0.

We can deploy multiple Azure Container Apps instances in a single environment. Physically, they will be in the same virtual network and isolated from the external network.

To enable monitoring, we have to create a Log Analytics workspace for each environment shared with Azure Container Apps.

In the next section, we will install Kubernetes locally and orchestrate microservices and multi-container applications.

Orchestrating microservices and multi-container applications

In the last chapter, we prepared the environment for Docker and installed Docker Desktop to deploy containers locally. Now, we will enable Kubernetes using the GUI tool of Docker Desktop by following these steps:

1. Open Docker Desktop and click on the settings icon in the top-right of the screen:

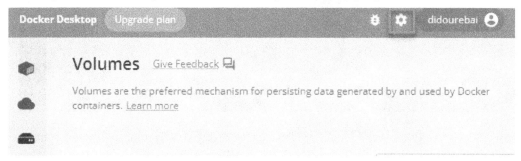

Figure 9.6 – Docker Desktop settings

2. Select **Kubernetes** from the left panel and check **Enable Kubernetes**, then click on **Apply & Restart**.

3. A new dialog window will be displayed to confirm the Kubernetes cluster installation. Select **Install**:

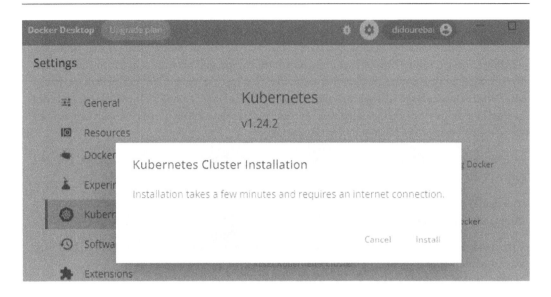

Figure 9.7 – Kubernetes Cluster Installation dialog

4. Let's run the `kubectl` command to display the Pods that are part of the Kubernetes system:

    kubectl get pods -n kube-system

    The result is presented in the following figure:

```
PS C:\WINDOWS\system32> kubectl get pods -n kube-system
NAME                                     READY   STATUS    RESTARTS   AGE
coredns-6d4b75cb6d-8w2n5                 1/1     Running   0          6m42s
coredns-6d4b75cb6d-wf42j                 1/1     Running   0          6m42s
etcd-docker-desktop                      1/1     Running   0          6m40s
kube-apiserver-docker-desktop            1/1     Running   0          6m44s
kube-controller-manager-docker-desktop   1/1     Running   0          6m41s
kube-proxy-c7x9t                         1/1     Running   0          6m43s
kube-scheduler-docker-desktop            1/1     Running   0          6m44s
storage-provisioner                      1/1     Running   0          6m35s
vpnkit-controller                        1/1     Running   0          6m35s
```

Figure 9.8 – List of Pods in the Kubernetes system

5. Next, we will install Kubernetes Dashboard because it isn't set up by default. We can deploy the Kubernetes Dashboard application using the `kubectl` command-line tool. Run the following command:

    **kubectl apply -f https://raw.githubusercontent.com/
    kubernetes/dashboard/v2.6.1/aio/deploy/recommended.yaml**

We get the following result:

```
Administrateur : Windows PowerShell
PS C:\> kubectl apply -f https://raw.githubusercontent.com/kubernetes/dashboard/v2.6.1/aio/deploy/recommended.yaml
namespace/kubernetes-dashboard created
serviceaccount/kubernetes-dashboard created
service/kubernetes-dashboard created
secret/kubernetes-dashboard-certs created
secret/kubernetes-dashboard-csrf created
secret/kubernetes-dashboard-key-holder created
configmap/kubernetes-dashboard-settings created
role.rbac.authorization.k8s.io/kubernetes-dashboard created
clusterrole.rbac.authorization.k8s.io/kubernetes-dashboard created
rolebinding.rbac.authorization.k8s.io/kubernetes-dashboard created
clusterrolebinding.rbac.authorization.k8s.io/kubernetes-dashboard created
deployment.apps/kubernetes-dashboard created
service/dashboard-metrics-scraper created
deployment.apps/dashboard-metrics-scraper created
PS C:\>
```

Figure 9.9 – Installing and deploying Kubernetes Dashboard

6. We will enable access to the Dashboard using the `kubectl` command-line tool. Run the following command:

```
kubectl proxy
```

The dashboard should be available at this link: `http://localhost:8001/api/v1/namespaces/kubernetes-dashboard/services/https:kubernetes-dashboard:/proxy/#/login`

7. A pop-up window will appear; click on **Skip**. We will get the following page presenting Kubernetes Dashboard:

Figure 9.10 – Kubernetes Dashboard

We are now ready to deploy multiple containers and use Kubernetes to orchestrate between them. We can deploy a microservices-based application and ensure communication between the different containers using Kubernetes.

Summary

In this chapter, we compared container orchestration and Docker containers. Next, we explored the basic architecture and different components of Kubernetes and considered the different cloud services in Azure – Azure Kubernetes Service and Azure Container Apps. Lastly, we prepared a local environment to orchestrate different microservices and a multi-container application.

In the next chapter, we will set up a Kubernetes cluster and deploy it on Azure Kubernetes Service using both the Azure portal and Azure CLI.

Further reading

*   Orchestrate microservices and multi-container applications for high scalability and availability: `https://learn.microsoft.com/en-us/dotnet/architecture/microservices/architect-microservice-container-applications/scalable-available-multi-container-microservice-applications`

10

Setting Up a Kubernetes Cluster on AKS

This chapter will cover setting up a Kubernetes cluster on **Azure Kubernetes Service** (**AKS**). We will create a new AKS cluster using the Azure CLI and Azure portal. Then, we will deploy an application on AKS using multiple methods: the portal, the Azure CLI, **Azure Resource Manager** (**ARM**) templates, and Azure DevOps Starter.

In this chapter, we're going to cover the following main topics:

- Exercise 1 – creating an AKS cluster using the Azure CLI
- Exercise 2 – creating an AKS cluster using the Azure portal
- Exercise 3 – deploying an AKS cluster and running an application using the Azure CLI
- Exercise 4 – deploying an AKS cluster using an ARM template
- Exercise 5 – deploying an AKS cluster using Azure DevOps Starter
- Exercise 6 – debugging your application using Bridge to Kubernetes

Exercise 1 – creating an AKS cluster using the Azure CLI

In this section, we will create a new AKS cluster using the Azure CLI.

You can open PowerShell and connect to your Azure account using your credentials. If you use the `az login` command line, you will be redirected to the web interface of Azure to log in to your account. Once you have logged in, you will get JSON output that includes all the information related to your subscription:

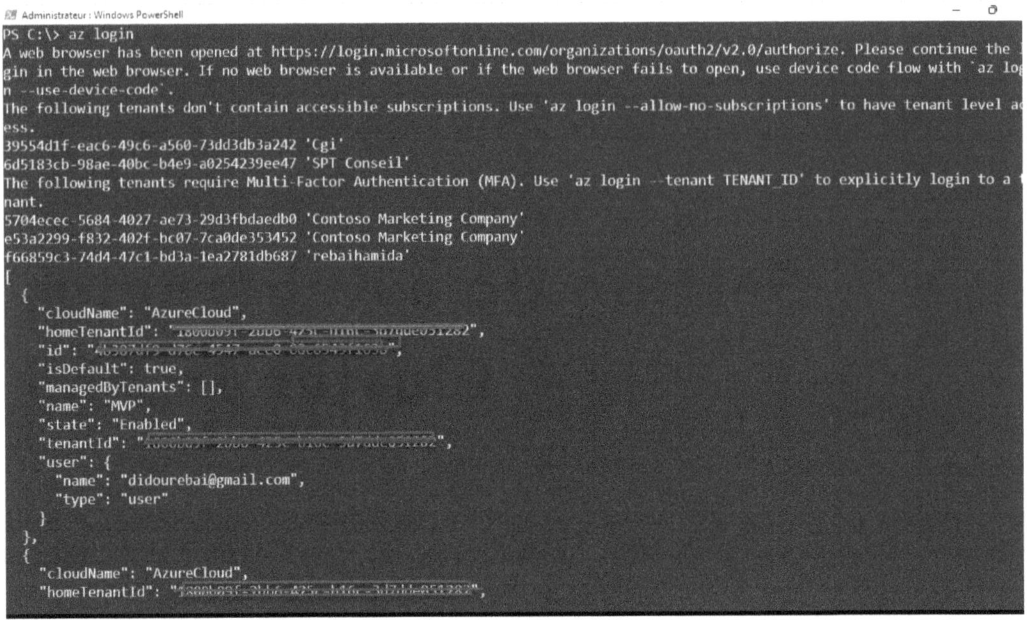

Figure 10.1 – Creating an AKS cluster using the Azure CLI

You can use the web interface of Azure Cloud Shell by going to `https://shell.azure.com/`.

Follow these steps to create a new AKS cluster:

1. Create a new resource group.

    We will start by creating a new resource group that will include our AKS cluster with the different resources that will be created with the cluster. We will use the `az group create` command, as shown:

    ```
    az group create --name healthcareRG --location eastus
    ```

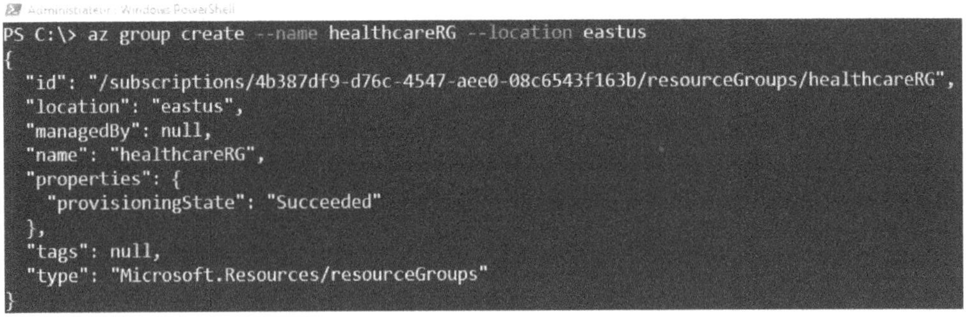

Figure 10.2 – Creating a new resource group

Once created, we will get JSON output that includes the different settings of the resource group.

2. Create an AKS cluster.

    We will use the `az aks create` command to create the cluster with a single node and attach an Azure Container Registry (ACR) using `--attach-acr`, as shown:

    ```
    az aks create --resource-group healthcareRG --name
    myAKSCluster --node-count 1 --attach-acr v --generate-
    ssh-keys
    ```

    This is what the output looks like:

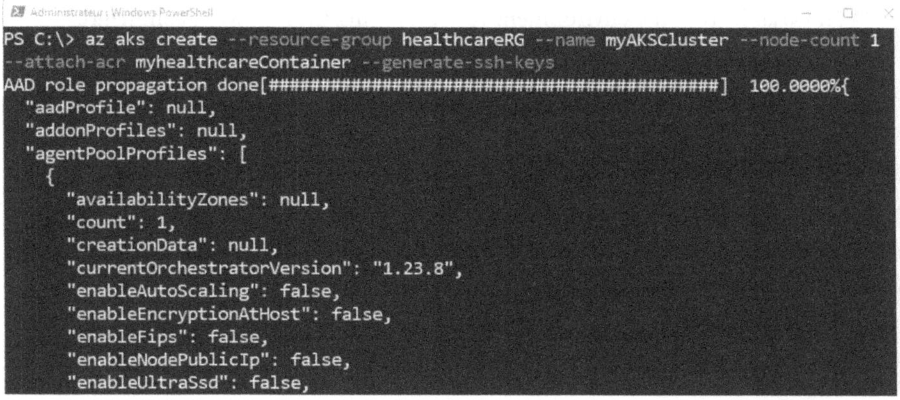

Figure 10.3 – Creating an AKS cluster

Once created, we will get JSON output that includes the different settings of the AKS cluster. If you have not configured ACR integration, you can update your cluster as shown:

```
az aks update -n myAKSCluster -g healthcareRG --attach-
acr <acr-resource-id>
```

3. Install the `kubectl` utility.

    We will connect to the AKS cluster, but we need to install the `kubectl` utility before doing so by using the following command:

    ```
    az aks install-cli
    ```

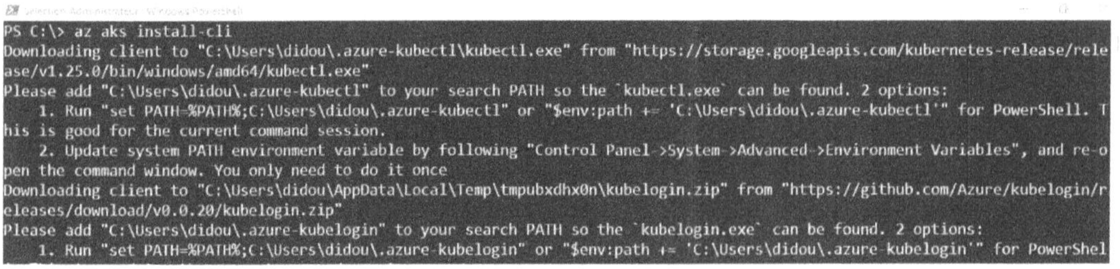

Figure 10.4 – Installing the kubectl utility

As mentioned in the results of the installation, we need to set the variable's path. We can set the PATH variable from the environment variables using the following command:

```
set PATH=%PATH%;C:\Users\youusername\.azure-kubectl
set PATH=%PATH%;C:\Users\yourusername\.azure-kubelogin
```

4. Connect to the AKS cluster.

We will need the **kube** configuration file to connect to our AKS cluster. We will use az aks get-credentials to get the configuration file, and we will use the following command to extract the kube configuration file to the local user's home directory:

```
az aks get-credentials --resource-group healthcareRG
--name myAKSCluster
```

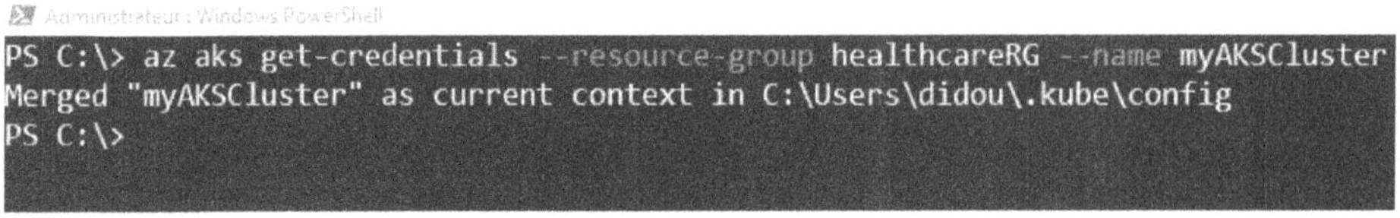

Figure 10.5 – Connecting to the AKS cluster

To verify the connection to the AKS cluster, we will use kubectl get nodes. This command will display all the nodes in the cluster:

Figure 10.6 – Getting all nodes in the AKS cluster – single node

We created a single node at the beginning, so the result will be similar to what's shown in *Figure 10.6*.

If we need to add more nodes, we can scale the cluster nodes using this command:

```
az aks scale --resource-group healthcareRG --name
myAKSCluster --node-count 2
```

We can also check the number of nodes again using `kubectl get nodes`:

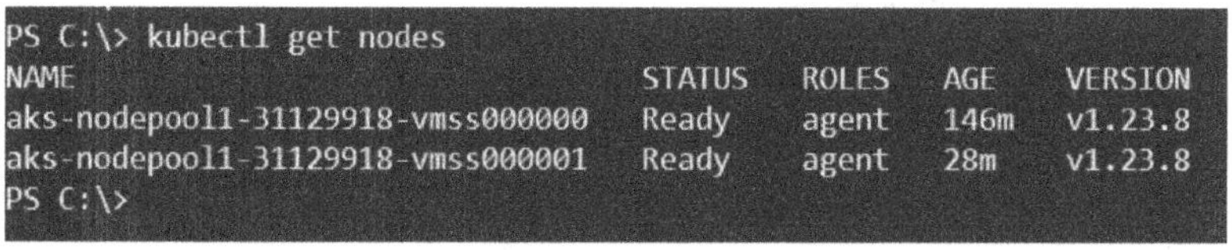

```
PS C:\> kubectl get nodes
NAME                                    STATUS   ROLES   AGE    VERSION
aks-nodepool1-31129918-vmss000000       Ready    agent   146m   v1.23.8
aks-nodepool1-31129918-vmss000001       Ready    agent   28m    v1.23.8
PS C:\>
```

Figure 10.7 – Getting all nodes in the AKS cluster – two nodes

When we create a new cluster, we can add more properties. For example, we can enable the cluster autoscaler to meet our application requirements on AKS. This is an example of updating an existing AKS cluster by enabling the cluster autoscaler:

```
az aks update --resource-group healthcareRG --name
myAKSCluster --enable-cluster-autoscaler --min-count 1
--max-count 3
```

After a few minutes, the cluster will be updated and configured with the cluster autoscaler settings.

In the next section, we will create our AKS cluster using the Azure portal.

Exercise 2 – creating an AKS cluster using the Azure portal

To create a new AKS cluster with the Azure portal, follow these steps:

1. Sign in to the Azure portal. Select **Create a resource**, then **Containers | Kubernetes Service**.

2. On the **Basics** tab, which is shown in the following figure, we will add the following settings:

Figure 10.8 – Create Kubernetes cluster – the Basics tab

To create a Kubernetes cluster, we must set up the following basic information for our AKS cluster:

- **Project details**: Select the Azure subscription. We can also create or select a resource group.

- **Cluster details**: We can choose a cluster preset configuration to customize the Kubernetes cluster. The options are **Standard**, **Dev/Test**, **Cost-optimized**, **Batch processing**, and **Hardened access**. We will select **Standard**. Fill in the **Kubernetes cluster name** field and choose a region. In the specified region, our master node will be created. Based on the region selected, we will select the availability zones. We will select the default version of Kubernetes and 99.95% or 99.5% API server availability.

- Under **Primary node pool**, select the node size. We will choose the default value, which is **Standard DS2 v2**, but you can select any size. It is recommended to select **Autoscale** for the scale method.

If you select **Autoscale**, you need to add the node count range (the minimum and the maximum number of nodes):

Figure 10.9 – Autoscale scale method and number of nodes

If you select **Manual**, you need to define the number of nodes that will be used by the cluster:

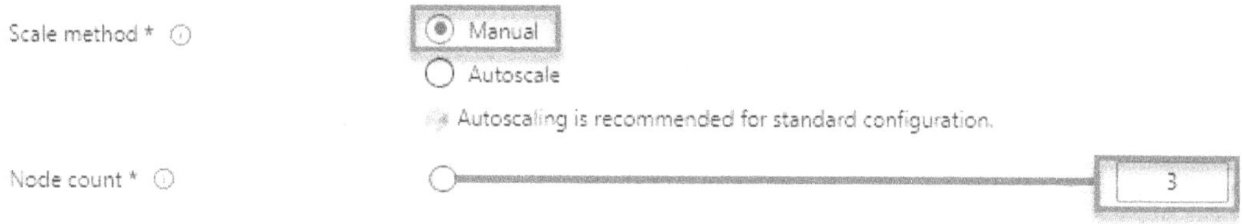

Scale method * ⓘ   ● Manual
                            ○ Autoscale
                            ⊙ Autoscaling is recommended for standard configuration.

Node count * ⓘ   ○━━━━━━━━━━━━━━━━━━━━━━━━━━━━━━━━ 3

Figure 10.10 – Manual scale method and the node count

In the **Node pools** tab, we have a default node pool, but we can add another node by selecting **Add node pool**. AKS groups nodes with the same configuration into node pools. Node pools contain the underlying virtual machines whose applications will run. If we need to deploy containers or burst containers to the nodes backed by serverless Azure Container Instances, we can enable virtual nodes. This can provide fast burst scaling options beyond your defined cluster size. A virtual node is a type of serverless container instance. Since we will be creating worker nodes as virtual machines, we will not enable this option:

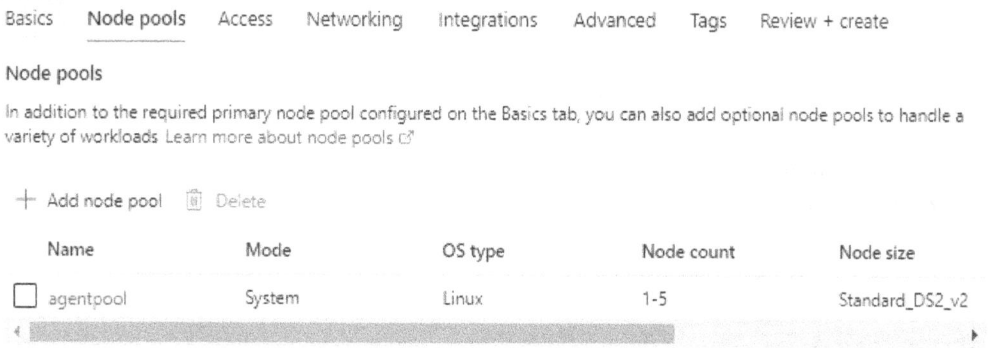

Basics   **Node pools**   Access   Networking   Integrations   Advanced   Tags   Review + create

Node pools

In addition to the required primary node pool configured on the Basics tab, you can also add optional node pools to handle a variety of workloads Learn more about node pools ⊡

+ Add node pool   🗑 Delete

Name	Mode	OS type	Node count	Node size
☐ agentpool	System	Linux	1-5	Standard_DS2_v2

Figure 10.11 – Create Kubernetes cluster – the Node pools tab

3. In the **Access** tab, we need to define the authentication and authorization method. We can use the default option, which is to use the local accounts with Kubernetes **Role-Based Access Control** (**RBAC**), use Azure **Active Directory** (**AD**) with Kubernetes RBAC for authorization, or use Azure AD authentication with Azure RBAC to perform authorization checks on the cluster:

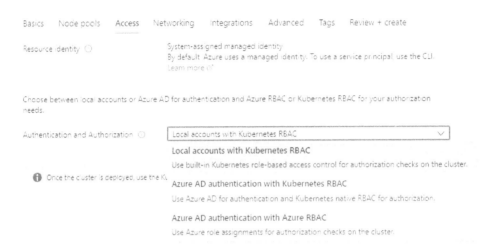

Figure 10.12 – Create Kubernetes cluster – the Access tab – Authentication and Authorization

4. In the **Networking** tab, we must fill in the network settings for our cluster. You can keep the default values or make some changes according to your needs:

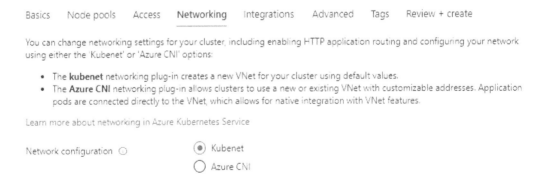

Figure 10.13 – Network configuration in the Networking tab

5. In the **Integrations** tab, we will select the container registry to be able to connect to it and enable seamless deployment from the private image registry. If we are creating AKS before ACR, we can click on **Create new** to create a new container registry:

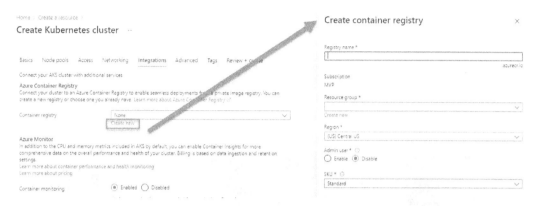

Figure 10.14 – Creating a new container registry in AKS cluster creation

In this tab, we have options for monitoring configuration. Here, we can enable **Container monitoring**, which allows us to monitor our container. We need to add a Log Analytics workspace to store monitoring data. We can also check **Use managed identity**, which is a secure authentication model where monitoring agents can use the cluster's managed identity to send all data to Log Analytics or any configured endpoints, and can enable Azure Policy to enforce more security on the AKS cluster:

Connect your AKS cluster with additional services

Azure Container Registry
Connect your cluster to an Azure Container Registry to enable seamless deployments from a private image registry. You can create a new registry or choose one you already have. Learn more about Azure Container Registry ↗

Container registry	myhealthcareContainer ⌄

Create new

Azure Monitor
In addition to the CPU and memory metrics included in AKS by default, you can enable Container Insights for more comprehensive data on the overall performance and health of your cluster. Billing is based on data ingestion and retention settings.
Learn more about container performance and health monitoring
Learn more about pricing

Container monitoring	◉ Enabled ○ Disabled

Azure monitor is recommended for standard configuration.

Log Analytics workspace ⓘ	DefaultWorkspace-4b387df9-d76c-4547-aee0-08c6543f163b-CCAN ⌄

Create new

Use managed identity (preview) ⓘ   ☐

Figure 10.15 – The Integrations tab – selecting a container registry

6. Click on **Review + Create**. Once the validation has passed, select **Create**.

Once created, we can connect to the AKS cluster using two methods: Cloud Shell or the Azure CLI.

At the top of the Azure portal, select the Cloud Shell icon:

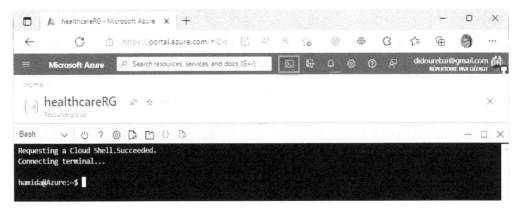

Figure 10.16 – Cloud Shell in the Azure portal

Run the following command:

```
az aks get-credentials --resource-group healthcareRG
--name myAKSCluster
```

To get the nodes running in your cluster, run the following command:

```
kubectl get nodes
```

The result is presented in the following figure:

```
hamida@Azure:~$ az aks get-credentials --resource-group healthcareRG --name myAKSCluster1
Merged "myAKSCluster1" as current context in /home/hamida/.kube/config
hamida@Azure:~$ kubectl get nodes
NAME                                STATUS   ROLES   AGE     VERSION
aks-agentpool-37518842-vmss000000   Ready    agent   3m33s   v1.23.8
aks-agentpool-37518842-vmss000001   Ready    agent   3m33s   v1.23.8
aks-agentpool-37518842-vmss000002   Ready    agent   3m32s   v1.23.8
```

Figure 10.17 – The number of nodes in an AKS cluster

We can find these commands when we open the overview of the AKS cluster in the Azure portal and select **Connect**:

Figure 10.18 – Connecting to an AKS cluster

We can update the cluster configuration in the Azure portal. We can also add a new node pool if needed. In the next section, we will deploy an application in the cluster using the Azure CLI.

Exercise 3 – deploying an AKS cluster and running an application using the Azure CLI

In the previous section, we created an AKS cluster where we will deploy our application.

Deployments play a critical role in the Kubernetes environment because they are the mechanism for creating Pods that run the containers hosting our code.

If we go back to the Kubernetes architecture, we have our master nodes, a database storing all the information, the controllers, which allow us to interact with the nodes, and the API that we interact with using `kubectl`. With Deployments, we can send our standard `kubectl` commands to the API and the controllers will arrange what happens on the nodes. We can create one or more Pods per node. Knowing that a node is a virtual machine, the goal of the Deployments is to deploy the Pods that will run in our containers. A Deployment relies on a **ReplicaSet**. A ReplicaSet is where we set the number of Pods. For example, if we give the ReplicaSet a value of 3, we will have three Pods.

Then, the Pod template defined in the Deployment determines what container or containers to run inside of that Pod. Pods need to talk to each other. So, in Kubernetes, we also have Services, and then we can have external storage, configuration data, such as ConfigMaps, and any Secrets for sensitive data.

A ReplicaSet is a declarative way to manage Pods. ReplicaSets verify that the number of Pods we requested is available. Suppose that a Pod is out of service because of some issues; the ReplicaSet will automatically bring a new one to replace it. A Deployment is a higher-level, declarative way to manage Pods, but it uses ReplicaSets. Deploying Pods in conjunction with a ReplicaSet is like conducting an orchestra, with different members of the orchestra becoming different Pods.

We haven't created a deployment yet, so there are no Pods. Say we use the following command:

```
kubectl get pods
```

The result will be as follows:

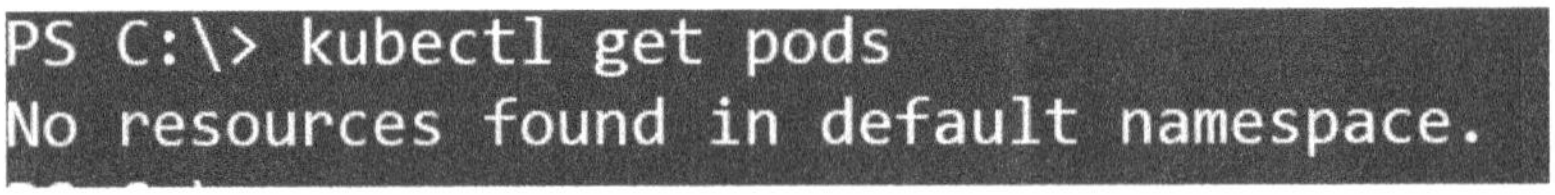

```
PS C:\> kubectl get pods
No resources found in default namespace.
```

Figure 10.19 – Getting Pods in an AKS cluster

We can create a deployment using the ACR where we pushed our application using the following command:

```
kubectl create deployment healthcare-application
--image=myhealthcarecontainer.azurecr.io/
healthcaresolutiondashboard:latest
```

We can expose the deployment on port 5000 using the following command line:

```
kubectl expose deployment healthcare-application
--type=LoadBalancer --port 5000 --target-port 5000
```

Then, we can run kubectl get service to find the external IP so we can try it out:

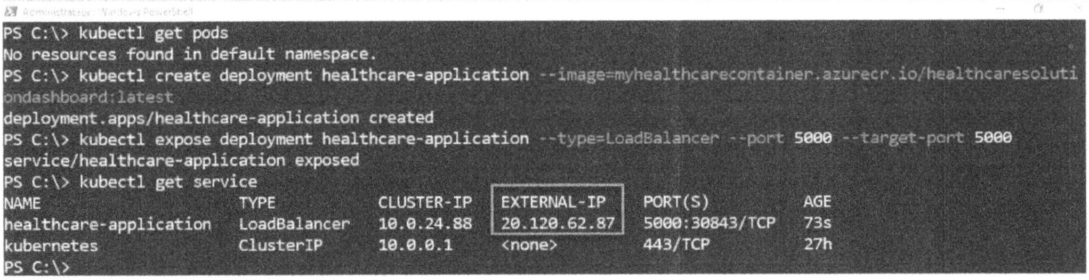

```
PS C:\> kubectl get pods
No resources found in default namespace.
PS C:\> kubectl create deployment healthcare-application --image=myhealthcarecontainer.azurecr.io/healthcaresoluti
ondashboard:latest
deployment.apps/healthcare-application created
PS C:\> kubectl expose deployment healthcare-application --type=LoadBalancer --port 5000 --target-port 5000
service/healthcare-application exposed
PS C:\> kubectl get service
NAME                     TYPE           CLUSTER-IP    EXTERNAL-IP     PORT(S)          AGE
healthcare-application   LoadBalancer   10.0.24.88    20.120.62.87    5000:30843/TCP   73s
kubernetes               ClusterIP      10.0.0.1      <none>          443/TCP          27h
PS C:\>
```

Figure 10.20 – Creating a deployment in an AKS cluster

You can open your application using this IP with port 5000: 20.120.62.87:5000.

If we need to deploy more than one application, we can create a manifest deployment file. Here, we can define things such as the Service or a deployment. When we define a deployment, we can select the strategy we want the cluster to use when creating a new deployment, which means that we can create all objects needed to run the application. This manifest is a file that includes the Kubernetes Deployments.

Before creating a new YAML file, we need to understand the difference between a deployment and a service. In a deployment, we create Pods from a template to maintain a running set of Pods. We manage a set of identical Pods in a Kubernetes object, which is the deployment. If we don't add a deployment block to the YAML file, we will have to create, update, and delete the Pods manually. A service is used to authorize network access to a set of Pods. We will select the Pods to operate by using labels and label sectors.

Here, we will create an initial Deployment and add some Pods to the cluster. Let's look at the steps:

1. The first thing we must typically do to create a Deployment, ReplicaSet, and then Pods running containers is to find some YAML. We will add a YAML file and use the `kubectl` command to interact with the master API. A ReplicaSet can trigger the controllers and scheduler to schedule one or more Pods on the different worker nodes in our cluster:

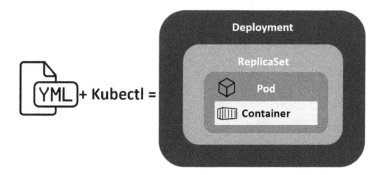

Figure 10.21 – Creating a deployment

2. The first thing you're going to have at the top is the API version. We will use `app/v1` as the Kubernetes API version. We also have `kind`, which presents the object that needs to be created. The value of `kind` is `Deployment`.

3. Then, you can have metadata, such as what you want to call this **Deployment** – in our case, `booking-ui-deployment`. The important part of a Deployment is the spec and the selector. The spec allows you to define your Pod template and have a selector that matches that Pod template.

4. Now, in addition to that, we have to define the container or containers to run. In our case, we have only one container. We're going to call our container `myhealthcarecontainer`, and we're going to take our image from ACR using `myhealthcarecontainer.azurecr.io/healthcaresolutiondashboard:latest`.

    This template will then be used to create the Pod. We will leave the default port of `80` for the container.

We can define the number of copies of each Pod we need to create using replicas. It is not mandatory to add a replica if we need to create one Pod. That's the basics of a Deployment YAML file, which looks as follows:

```
apiVersion: apps/v1
kind: Deployment
metadata:
  name: booking-frontend-deployment
spec:
  selector:
    matchLabels:
      app: booking-ui-pod
  replicas: 3
  template:
    metadata:
      labels:
        app: booking-ui-pod
    spec:
      containers:
      - name: myhealthcarecontainer
        image: myhealthcarecontainer.azurecr.io/
healthcaresolutiondashboard:latest
        ports:
        - containerPort: 80
```

How can we move this YAML file into the master node to schedule these Pods on worker nodes?

We can create an empty file using the `vi` or `nano` command line. To be able to use it in Azure Cloud Shell, we can use the terminal in Visual Studio to deploy the file:

```
$ vi healthcare-deployment.yaml
```

Before copying the previous YAML content, we will make some updates by adding the service part. `apiVersion` will be `v1` and `kind` will be `Service`. We will add the metadata, which includes the name and the spec. Inside the spec, we will identify the port that will be exposed by the service. `targetport` is the port on which the service will target the container and `nodePort` is the port in which all the services will be exposed in all the nodes inside the cluster:

```
---
apiVersion: v1
kind: Service
metadata:
  name: booking-frontend-service
spec:
  type: NodePort
  selector:
    app: booking-ui-pod
  ports:
    - name: "http"
      port: 80
      targetPort: 80
      nodePort: 30180
      targetPort: 80
      nodePort: 30180
```

You can add more than one Deployment and Service in the same file; it depends on your application. Kubernetes will orchestrate these containers. We can add a single file that includes all the deployments.

We will use the kubectl create command:

```
$ kubectl create -f healthcare-deployment.yaml --save-config
```

We will add –save-config to store the current properties that we're initially creating on the Deployment in the resource's annotations.

We will use kubectl, apply -f, and use the YAML. If that resource doesn't exist, it's like performing a create operation. If it does exist and we've changed something, such as the image version inside of the Pod template of that Deployment, these changes will be applied, which will cause the Pods to get updated with the new updated container image.

kubectl scales the deployment and then gives it a name:

```
$ kubectl apply -f healthcare-deployment.yaml
```

The result will be as follows:

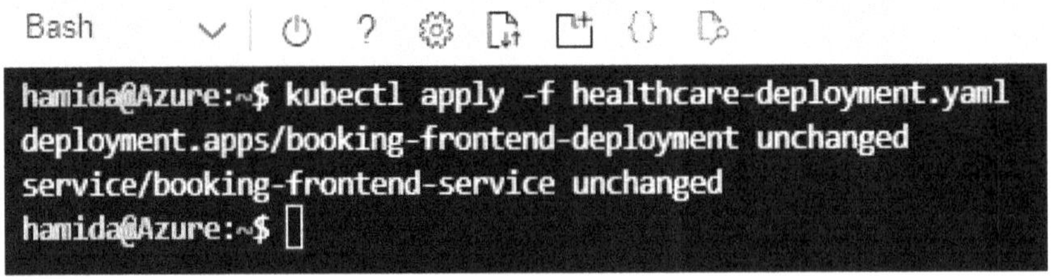

Figure 10.22 – Deploying an application using kubectl apply

We will check our deployment using this command:

```
$ kubectl get deployments
```

We can assign a valid public IP address to the service using this command:

```
$ kubectl get services
```

The result is presented in the following figure. We will use the public IP to display our application:

Figure 10.23 – Getting deployed services

If we need to scale the Deployment of Pods to 5, for example, we can use this command:

```
kubectl scale deployment booking-frontend-deployment
--replicas=5
```

Alternatively, we can scale using the deployment file like so:

```
kubectl scale -f healthcare-deployment.yaml --replicas=5
```

The application is using a database, and the database has been deployed on a container. We need to deploy the database. We will set up a deployment as before with one Pod and one container. The configuration YAML file is similar to this:

```
apiVersion: apps/v1
kind: Deployment
```

```
metadata:
  name: healthcare-db-deployment
spec:
  replicas: 1
  selector:
    matchLabels:
      app: healthcaredb
  template:
    metadata:
      labels:
        app: healthcaredb
    spec:
      terminationGracePeriodSeconds: 30
      hostname: mssqlinst
      securityContext:
        fsGroup: 10001
      containers:
      - name: mssql
        image: mcr.microsoft.com/mssql/server:2019-latest
        resources:
          requests:
            memory: "2G"
            cpu: "2000m"
          limits:
            memory: "2G"
            cpu: «2000m»
        ports:
        - containerPort: 1433
        env:
        - name: MSSQL_PID
          value: "Developer"
        - name: ACCEPT_EULA
          value: "Y"
        - name: MSSQL_SA_PASSWORD
          valueFrom:
            secretKeyRef:
```

```
            name: mssql
            key: MSSQL_SA_PASSWORD
        volumeMounts:
        - name: mssqldb
          mountPath: /var/opt/mssql
      volumes:
      - name: mssqldb
        persistentVolumeClaim:
          claimName: mssql-data
---
apiVersion: v1
kind: Service
metadata:
  name: mssql-deployment
spec:
  selector:
    app: mssql
  ports:
    - protocol: TCP
      port: 1433
      targetPort: 1433
  type: LoadBalancer
```

We will follow the same steps to deploy this file. We can save it in a file and use `kubectl apply` to deploy it. Don't forget to make changes to the credentials to access the database.

We can also deploy an AKS cluster using an ARM template. We will learn how to do that in the next section.

Exercise 4 – deploying an AKS cluster using an ARM template

An ARM template is a JSON file that defines the different elements related to the infrastructure and configuration for a project.

The template uses a declarative syntax to describe a deployment without writing on a command line to create the deployment of a cluster.

To deploy a new cluster, we can follow this link, which includes a template for creating a new cluster in AKS: `https://github.com/Azure/azure-quickstart-templates/tree/master/quickstarts/microsoft.kubernetes/aks`.

When we click on **Deploy to Azure**, we will be redirected to a new page where we can create a new AKS cluster. We can also edit the template or the parameters, or visualize our cluster and the different dependencies:

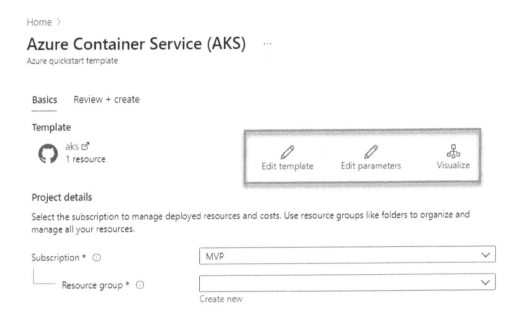

Figure 10.24 – Creating an AKS cluster from an ARM template

We will add the different information needed; it is similar to the classic process of creating a new AKS resource. However, we need to define the SSH public key source. We can create a new key by adding a key pair name while creating the AKS cluster. We can also use an existing key stored in Azure:

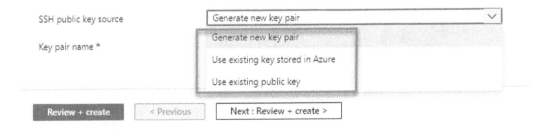

Figure 10.25 – SSH public key source

To create a new SSH key pair, we can use the `ssh-keygen` command, as follows:

```
ssh-keygen -t rsa -b 4096
```

Once you've added the different information and it's been validated, click on **Review + create**. Select **Create** and your AKS cluster will be created.

We can use a **Bicep** file to deploy an AKS cluster similar to an ARM template and run multiple container applications. Note that Bicep is a **domain-specific language** (**DSL**) used as a declarative syntax to deploy Azure resources.

You can follow this documentation link to deploy AKS using a Bicep file: `https://learn.microsoft.com/en-us/azure/aks/learn/quick-kubernetes-deploy-bicep?tabs=azure-cli%2CCLI`.

We can also use Azure DevOps Starter to deploy an AKS cluster. We will look at that in the next section.

Exercise 5 – deploying an AKS cluster using Azure DevOps Starter

To deploy an AKS cluster using Azure DevOps Starter, we will open the Azure portal and search for `DevOps Starter`:

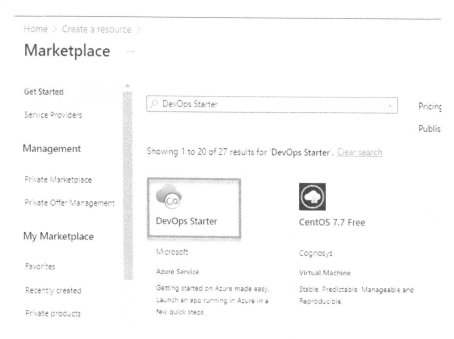

Figure 10.26 – Azure DevOps Starter

Select **Create**. The first step consists of defining whether we will start a new application or select existing application source code from GitHub or Azure DevOps.

If we start a new application, we must click **here** to change the settings. We can select **GitHub** or **Azure DevOps**. We can also select our language: **.NET**, **Node.JS**, **PHP**, **Java**, **Static Website**, **Python**, **Ruby**, or **Go**.

If we select **.NET**, we can click on **Next: Framework**. We have two options: **ASP.NET** or **ASP.NET Core**. We will select **ASP.NET Core**. After clicking on **Next: Service** to select an Azure service to deploy the application, we will select **Kubernetes Service**. After that, click on **Next: Create**:

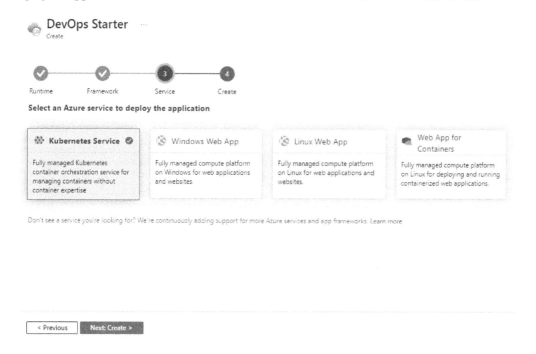

Figure 10.27 – Selecting Kubernetes Service to deploy an
application using Azure DevOps Starter (option 1)

If you select **Azure DevOps**, you will have more options to select from, such as **Service Fabric Cluster**, **Virtual machine**, and **Function App**:

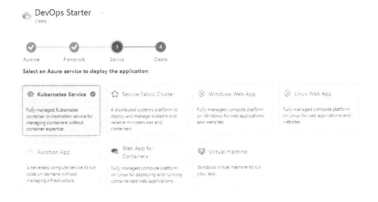

Figure 10.28 – Selecting Kubernetes Service to deploy an
application using Azure DevOps Starter (option 2)

We will select **Authorize** to allow permission to access the **GitHub** account since, in the beginning, we selected **GitHub**:

Figure 10.29 – DevOps Starter – Select Repository and Subscription

If we select **Authorize**, we will confirm the authentication to GitHub and be taken to the page shown in the following figure. Add the information related to the organization, your repository, the cluster name, and the node size.

We can use an existing cluster if we have already created one using the Azure portal, the Azure CLI, Azure PowerShell, or an ARM template.

Confirm its creation by clicking on **Review + Create** after **Create**:

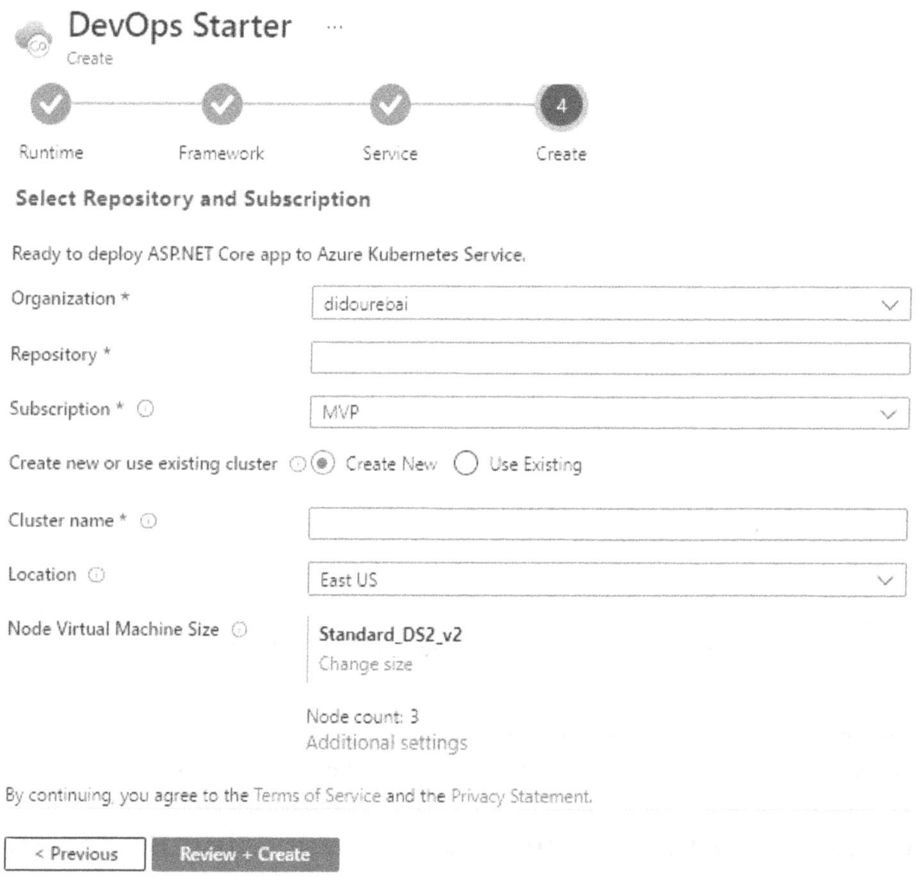

Figure 10.30 – GitHub repository – deploying an ASP.NET Core app to AKS

If we selected **Azure DevOps** at the beginning, we will be redirected to a page where we can add the project name and the different elements to deploy our cluster, such as the cluster's name, location, and node size:

Figure 10.31 – Azure DevOps Repository – deploying an ASP.NET Core app to AKS

Once your cluster has been deployed and you've configured your pipeline in GitHub or Azure DevOps, you will be able to commit your changes and deploy them automatically to Azure.

When we deploy our application in an AKS cluster and we get an issue, we need to debug using our development computer to be able to find the problem or exception. One way we can do that is with Bridge to Kubernetes. We will see how it works in the next section.

Exercise 6 – debugging your application using Bridge to Kubernetes

After deploying your application to an AKS cluster, one of the most common questions is, "*how can I debug my application?*" There are several ways to answer this question, and in this section, we will explore Bridge to Kubernetes.

Bridge to Kubernetes is an extension that we can add to Visual Studio 2022 or Visual Studio Code that allows us to run and debug our application in a development environment.

Bridge to Kubernetes in Visual Studio 2022

We will open our solution in Visual Studio 2022. Select **Extensions** at the top, then **Manage Extensions**. Search for `Bridge to Kubernetes` and then click on **Download** to add it:

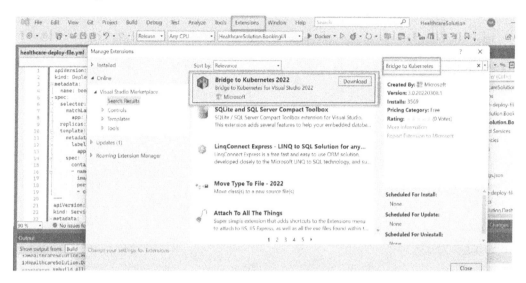

Figure 10.32 – Adding the Bridge to Kubernetes extension

Restart Visual Studio and, in the project, select **Bridge to Kubernetes** from the settings at the top:

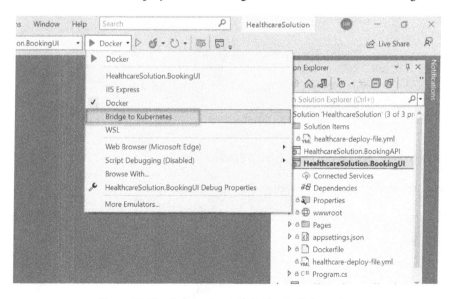

Figure 10.33 – Debugging with Bridge to Kubernetes

We will select the start button next to **Bridge to Kubernetes** and do the following in the dialog box to create a profile for Bridge to Kubernetes:

1. Select the cluster name where you deployed your application (Deployment and Service).

2. Select your namespace and service.

3. Select the same URL you used previously to launch your browser.

You can select **Enable routing isolation** so that cluster users are not affected by any change. In isolation mode, requests are forwarded to a copy of each affected service. Other traffic forwards normally. We will select **Save and debug**:

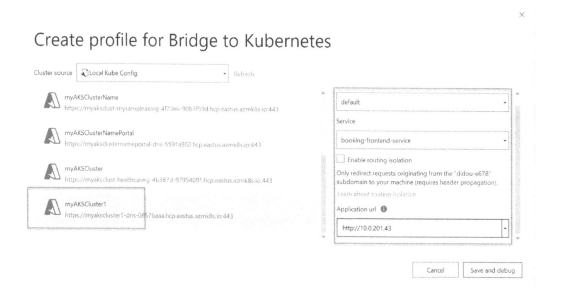

Figure 10.34 – Creating a profile for Bridge to Kubernetes

We can also add a breakpoint to debug our solution.

Bridge to Kubernetes in Visual Studio Code

Follow these steps to add Bridge to Kubernetes in Visual Studio Code:

1. Open Visual Studio Code. To add the **Bridge to Kubernetes** extension, select **Install**, then **Enable**:

Figure 10.35 – Adding the Bridge to Kubernetes extension in Visual Studio Code

2.   Select **Enable** and then use *Ctrl + Shift + P* to debug using Bridge to Kubernetes.

3.   Open the Command Palette (*Ctrl + Shift + P* or *Cmd + Shift + P* on a Mac) and run the **Bridge to Kubernetes: Configure** command to start the configuration process:

Figure 10.36 – Bridge to Kubernetes: Configure

4.   Select the Kubernetes service that needs to be redirected to the local machine:

Figure 10.37 – Selecting a service

5. Enter your local port number, such as 80 or 0, if traffic redirection is not needed and press *Enter* to confirm or *Escape* to cancel.

6. We can also create a new launch configuration. A bridge will add a dedicated launch configuration in `.vscode/launch.json` and we can configure Bridge to Kubernetes without a launch configuration. Press *Enter* to continue.

7. We will select whether we want to isolate the local version of the service from other developers. Select **Yes** or **No** after pressing *Enter*.

Bridge to Kubernetes is now configured. If we need to use it, we can open the Command Palette (*Ctrl + Shift + P*) again and select **Bridge to Kubernetes: Open Menu**, at which point we can add a breakpoint and start debugging.

Summary

In this chapter, we created a new cluster on AKS using the Azure portal, the Azure CLI, and ARM templates. After that, we deployed an AKS cluster and ran an application using the Azure CLI. We used Azure DevOps Starter to deploy a new application and create a new AKS cluster with a pipeline. In the end, we used Bridge to Kubernetes to debug an application in Visual Studio 2022 and Visual Studio Code.

In the next chapter, we will introduce Azure DevOps and GitHub.

Further reading

If you need more details about the topics that were covered in this chapter, you can refer to these documentation pages:

- `https://learn.microsoft.com/en-us/azure/aks/`
- `https://kubernetes.io/docs/home/`

Part 4 – Ensuring Continuous Integration and Continuous Deployment on Azure

In this final part of the book, we will focus on continuous integration and continuous deployment on Azure.

This part comprises the following chapters:

11

Introduction to Azure DevOps and GitHub

Strong demand for the rapid production of solutions while guaranteeing quality has made it necessary to switch to a project-based mindset and rethink the various processes that are usually used during the software development life cycle.

As DevOps-enabled software continues to take over the world at an accelerating pace, the challenges and struggles organizations face when implementing DevOps continue to be very real. These challenges can be overcome by working together to improve tools, processes, and knowledge, as well as by training employees.

This chapter will provide an introduction to Azure DevOps and GitHub.

In this chapter, we're going to cover the following main topics:

- What is DevOps?
- Exploring Azure DevOps
- Exploring GitHub
- Exercise – creating an Azure DevOps organization

What is DevOps?

Our first definition of DevOps comes from Gartner Research. One of their public definitions of DevOps is, *"DevOps represents a change in IT culture, focusing on rapid IT service delivery through the adoption of agile, lean practices in the context of a system-oriented approach."* So, culture comes before technology, and the goal is rapid delivery.

Another definition comes from Donovan Brown in *What is DevOps?*, where he writes, *"DevOps is the union of people, process, and products to enable the continuous delivery of value to our end users."* Follow this link for more information: https://www.youtube.com/watch?v=WW6xOjIPpr0.

DevOps is the practice of **development (Dev)** and **operations (Ops)** engineers collaborating through the entire software life cycle, from the design and development process to production support; it is the process and technology in application planning, development, delivery, and operations. DevOps replaces a model in which we have a team writing the code, a QA team working on testing the code, another team working on deploying the code, and yet another team that will operate it:

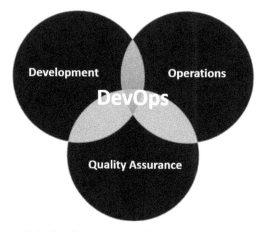

Figure 11.1 – DevOps stands for development and operations

DevOps is also characterized by operations personnel using the same techniques that developers use for their work on systems.

DevOps is used to ensure the release of better software. DevOps is a cultural practice, not just something to do with technical tools or workflows. DevOps produces better software faster by aligning the development, build, and deployment processes.

DevOps is a set of practices that work to automate and integrate the processes between software development and IT teams so that they can build, test, and release software faster and more reliably.

Like Agile or even Lean, DevOps is a broad concept with a single high-level definition. We can break it down into five main elements: values, principles, methods, practices, and tools:

- **Principles** include simplicity, automation, measurement, and management.
- **Methods** include operations such as Kanban with operations and visible Ops.
- **Practices** include continuous integration, continuous delivery, and continuous deployment.
- **Tools** support the different principles, methods, and/or practices.
- **Values** respond to the changes over following a plan. People over process over tools.

We need to understand the difference between continuous integration, continuous delivery, continuous deployment, continuous monitoring and Infrastructure as Code:

- **Continuous integration** (**CI**) and **continuous delivery/continuous deployment** (**CD**) are used in modern development practices and DevOps. What is the difference between them?

- **CI** encourages continuous code merging and testing, leading to the early detection of bugs. Other benefits include less time wasted dealing with merge issues and faster feedback to the development team.

- **CD** is an extension of CI. It is a semi-manual process that allows developers to deploy all changes to their customers with a simple click of a button. It also allows you to auto-deploy code changes to diverse environments (development, staging, testing, QA, production, and so on) so companies can quickly troubleshoot and fix bugs and respond to changing business needs.

- **Continuous deployment** emphasizes automating the entire process of deployment with every code change to production, which takes the process another step further than CD. This process is completely automated. Code should be deployed without breaking functionality that already works and is available to a large number of users.

- **Continuous monitoring** is the final phase of the DevOps life cycle and is about evaluating the entire cycle. The purpose of monitoring is to identify problem areas in the process, analyze team and user feedback to report inaccuracies, and improve product functionality.

- **Infrastructure as Code** (**IaC**) is an infrastructure management approach that enables continuous delivery and DevOps. We use scripts to automatically set the deployment environment (network, virtual machines, and so on) to the desired configuration, regardless of the initial state.

Let's explore the different phases of the DevOps application life cycle by looking at the following diagram:

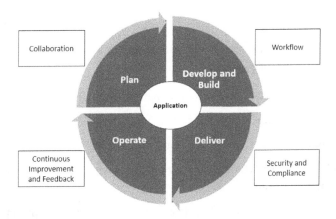

Figure 11.2 – DevOps application life cycle

In a DevOps journey, we start with a new opportunity to create a new solution. The first phase is the planning phase, where the DevOps team defines and describes the different features of the application. The team will plan, with an Agile methodology, to use the different DevOps practices, such as using Kanban boards, creating backlogs, and so on.

The next phase is development, where we can include all aspects of software code development. In this phase, the development environment is selected, and developers write code using version control systems such as Git to collaborate on code with other developers and code together in parallel. Here, the developers will test, review, and integrate the code. We can build code into different artifacts so that we can deploy it to multiple environments.

The delivery phase involves deploying applications to production environments, ideally via CD.

The operations phase involves maintaining, monitoring, and troubleshooting applications in production environments, including hybrid or public clouds such as Azure.

To manage the daily work and plan the best solutions for the future, it is essential to use tools when implementing DevOps. These tools level up your maturity: Azure DevOps and GitHub.

To help organizations define their work management tools and licensing strategy, we'll dive deeper into Azure DevOps and GitHub in the following sections.

Exploring Azure DevOps

To create software with a team, we need a set of tools that help plan the project, collaborate on code, run test cases, and build and deploy code efficiently. Software vendors often bundle these services together to improve the developer experience. This set of these tools is called a DevOps suite. Over the years, Microsoft has embraced the use of DevOps practices and developed tools to make DevOps workflows accessible to everyone.

Azure DevOps includes built-in functionality that you can access from a web browser using Azure servers or via an IDE client via the on-premises installable version. Depending on your business needs, you can use one or more of the following standalone services:

- **Azure Boards**, where the developer team can manage work efficiently.

- **Azure Repos**, where we host source code on Azure servers.

- **Azure Pipelines**, which ensures the hosting, building, and deployment of our code in the cloud.

- **Azure Test Plans** provides multiple tools for testing our app, including manual/exploratory testing and continuous testing.

- **Azure Artifacts** allows us to share different packages such as Maven, NuGet, npm, or any public or private sources. We can integrate these packages and share them using pipelines.

In the next section, we will introduce Azure Boards.

Azure Boards

Azure Boards is a customizable tool for managing software projects. It supports Agile, Scrum, and Kanban processes by default, and it's a hub for managing software projects with tools for planning, work assignment, and reporting. You can use Azure Boards to track work with Kanban boards, backlogs, team dashboards, and reports. Here, you can track project-related work, issues, and code errors. You can also create custom process templates and use them to create better and more customized business experiences.

We track work using standard work item types such as user stories, bugs, features, and epics.

You can customize these types or create your own. Each work item provides a standard set of system fields and controls, including instructions for adding and tracking comments, history, links, and attachments.

Work items are where you and your team describe the details of what work should occur in a software project. When you create a work item, you specify details about that work.

Azure Boards supports four process templates that map common software workflows. Basic and Agile are the most popular. Integration of Scrum with the **Capability Maturity Model** (**CMM**) is also supported. Each DevOps team can choose a process that fits their style. This process differs primarily in the work items provided to plan and track the work. When creating a new Azure DevOps project, choose to commit to a process that is standard across all project teams.

Azure Boards provides the following elements in terms of work items:

- **Boards**: Every project you create comes with a preconfigured Kanban board that's ideal for managing your workflow. Boards are highly customizable, so you can add the columns you need for every team and project. These boards support card customization, swim lanes, filtering, conditional formatting, and even **work item type** (**WIT**) restrictions. You can choose the type of work item to use when creating a project.

- **Backlogs**: These present work items as a list. The product backlog represents a project plan – that is, a roadmap of what the team will deliver. The backlog also provides a repository for all the information needed to track and share with your team. A backlog helps prioritize the different tasks and understand the relationships between them. We can customize them with drag-and-drop elements to adjust the order and quickly assign work to the next sprint.

- **Sprints**: These allow teams to create work items that can be completed together. Each sprint comes with a backlog, task board, burndown chart, and capacity planning view to help your team stay on track. The sprint backlog and task boards provide a filtered view of the work items your team has assigned to a specific iteration path or sprint. Sprints are defined for a project and selected by the team. You can drag and drop work from a backlog onto an iteration path and view that work in another sprint backlog.

- **Queries**: These let you precisely customize what you track and create KPIs that are easy to monitor. You can easily create new queries and pin them to your dashboard for quick monitoring and status. You can also list bugs, user stories, or other work items based on field criteria you specify using a query. You can then review these lists with your team, reorder work, and update work items in bulk. Apart from managed queries, semantic search tools offer some distinct and overlapping features that are worth investigating.

- **Delivery Plan**: This displays a calendar view presenting multiple backlogs, teams, and backlogs for different projects. Delivery Plan was integrated into the Azure marketplace extension:

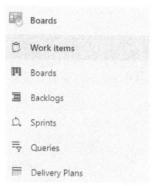

Figure 11.3 – Azure Boards – Work items

Creating a work item is straightforward. First, we must go to the **Boards** page, then select **Work items**. The page will display a list of all of your work items. From here, you can select the **New Work Item** button, where you'll see a list of potential work item types – that is, **Epic**, **Issue**, and **Task**:

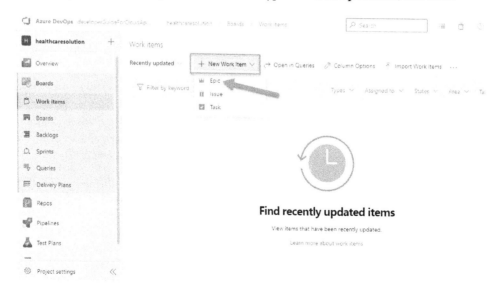

Figure 11.4 – New Work Item

Here, we will select the Scrum template for our example. Let's start by selecting **Epic** to find out more about how to use this template.

Epics are large packages of work with a common goal. They are considered groups of tasks with long timeframes. For example, planning and programming customer requirements can take weeks. Epics are typically split into small pieces based on customer and developer input.

We can display the group of Epics and we can filter by **Types** (**Epic**, **Issue**, or **Task**), **Assigned to**, **States** (to do, doing, or done), **Area**, and **Tags**.

Tasks track the actual work that needs to be done. A task is a small job that a developer needs to do. Tasks are in the sprint backlog and contain information about user stories and requirements. Task schedules are usually estimated on an hourly basis. In a test case, we have obstacles – that is, elements blocking progress on our project. Bugs are specific problems that you encounter in your application.

We can import work items via CSV files, which will add them to **Work items**.

Under **Backlogs**, we can add more than just issues by selecting **New Work Item**. Then, under **Sprints**, we can create a new sprint and add the tasks from the backlog.

You can create separate deliverable views and track dependencies across multiple teams using a calendar view that includes delivery schedules. Here, we can select a new plan, after which we can enter the name of the delivery plan, its description, and a team:

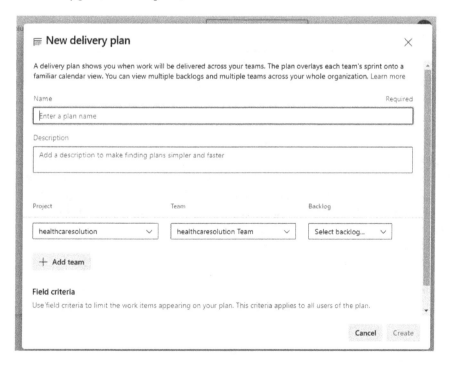

Figure 11.5 – Adding a new delivery plan

We can configure a dashboard using widgets or charts. To do so, we can select **Overview**, then **Dashboard**, and then **Add a widget**. We will see a list of widgets that we can add to our dashboard; these are based on the specific team. We can select more than one widget if we wish. When we're done, we can select **Done Editing**.

Azure Repos

A version control system is essential to any software project. Having frequent copies of the source code while each team member is working is essential to the health of the project. All modern version control systems support using a common repository location – that is, a location in a server or cloud where team members can access and update a shared code base.

Azure Repos is a location where we can store our repositories on Microsoft's cloud servers. Inside a DevOps organization, each project includes multiple repositories. When we build our project, a default repository is created. We have two options to host in Azure DevOps. Git is the default provider, but we can also select **Team Foundation Version Control** (**TFVC**). Both systems handle all common repository actions.

We can commit code changes, as well as manage files and folders in our projects. We can create new code branches and handle emergency conflicts too. Git has become the industry standard for version control. Git is a distributed system, which means each developer has a copy of the entire source repository on their dev machine.

Developers can commit a set of changes to their local machine and perform version control operations such as commit, view history, and view **DIFF** code (show changes between commits) without a network connection. Every developer can work with their Git repository on their local machine. When the time comes to collaborate with other users or work on software in teams, it makes sense to have a copy of the repository in a shared location. This is the `git remote` command. Then, different members of the team can work locally.

To sync with other teams, members must push their changes to the remote server when they're ready; alternatively, they can issue a pull request to the server so that their teammates can review them before pulling the changes from the local source code to the remote repository. Note that Azure Repos is a hosting service for Git remotes. GitHub is a Microsoft ecosystem and it also has a set of DevOps services, including the ability to serve as a Git repository host. It seems to be the most popular Git hosting service available. When talking about Git hosting, there is little difference between Azure DevOps and GitHub. Both are suitable hosting services for Git repositories. You can host on GitHub and use Azure DevOps for all the other services, or keep everything in Azure DevOps.

When we select a project, the entry point for our repositories can be found under **Repos**. Here, we can explore the different subsections: files, commits, pushes, branches, tags, and pull requests.

Under **Files**, we can find the different files in a project. Under **Commits**, we can see the history of these files, who committed them, and when. If we select any line in the commit list, we can display the details about that commit. We can also go back to the files and edit them; we can commit those files directly from the web portal. In the following figure, we can see the different changed files:

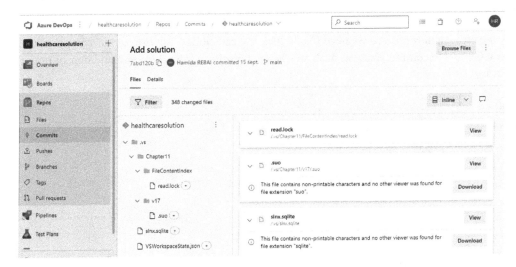

Figure 11.6 – Repos – Commits page

Under **Pushes**, we can display the pushes to one of the branches. We can look at all the pushes from a specific day.

Under **Branches**, we can display the different branches. The default branch is the `main` branch in this example, but we can have more than one:

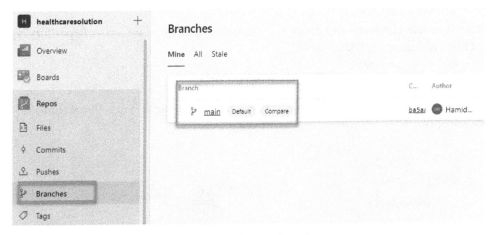

Figure 11.7 – Repos – Branches

Under **Tags**, we can add tags to a marked point in history that is relevant to the repository.

Under **Pull requests**, we can add a new pull request. By selecting **New pull request**, we can review the code and help the developers ensure it's of good quality before merging the branches.

Adding files

When we create a new project, we need to add files. We can clone a solution to our computer after we use a Git tool to commit changes, push an existing repository from the command line, or import a repository:

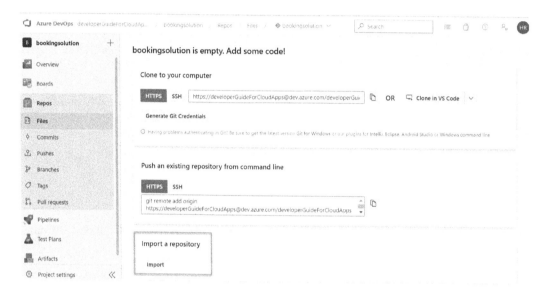

Figure 11.8 – Importing a repository in Azure DevOps

In this case, if we select **Import**, a new dialog box will be displayed, where we can set **Repository type** (**Git** or **TFVC**) and the clone URL.

Creating a repo branch using Visual Studio Code

We can clone a local copy by selecting **Clone in VS Code**. Upon doing so, a new Visual Studio Code instance will open:

Figure 11.9 – Clone in VS Code

We will start by cloning an existing repository. Then, we can get a local copy of the Git repository by clicking the **Copy to clipboard** button.

In Visual Studio Code, press *Ctrl + Shift + P* to display the Command Palette and execute the `Git: Clone` command:

Figure 11.10 – Git: Clone in Visual Studio Code

Add the URL to our repository and then press *Enter*, after which we can select a local path to clone our repository to. Finally, select **Open** to open the cloned repository:

Figure 11.11 – Clone from URL in Visual Studio Code

Now, we can add a new application using the .NET CLI in the terminal of Visual Studio Code:

```
dotnet new webapi -o DoctorAPIs
cd DoctorAPIs
dotnet add package Microsoft.EntityFrameworkCore.InMemory
code -r ../ DoctorAPIs
```

Then, we can use the `Git: Commit` command to commit the changes in the application. Alternatively, we can use the terminal and use the following command:

```
git commit -m'my first commit'
```

We can push to the branch master using `git push`. If we go to Azure DevOps and select **Repos Files**, we will find that a new repository has been added.

Creating a repo branch using Visual Studio 2022

At the bottom of Visual Studio 2022, select **Select Repository | Clone Repository…**:

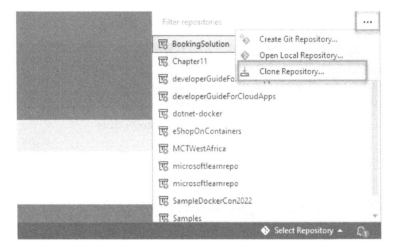

Figure 11.12 – Clone Repository… in Visual Studio 2022

Here, we can enter the Git repository URL and the path. Alternatively, we can browse a repository from Azure DevOps and select the repository to clone by selecting **Clone Repository…**.

Exploring GitHub

GitHub is one of the most popular open source project management tools available. Git is a version control system; GitHub is a website and a platform for people to share a Git repository online (a repository, or repo, is a snapshot of one version of your code).

To start using GitHub, we will follow these steps:

1. Create and configure a new organization.
2. Configure GitHub projects.
3. Create a new repository.

Let's get started.

Create and configure a new organization

When we create a new organization, there are no repositories associated with it. To create a new organization, perform the following steps:

1. Click on the profile picture in the top-right corner of any page. Then, click on **Settings**.

2. In the **Access** section, select **Organizations**.

3. In the **Organizations** section, select **New organization**.

Configure GitHub projects

To get started with **GitHub** projects, we need to create an organization or a user project. To create an organization project, follow these steps:

1. In GitHub, navigate to the main page of your organization.

2. Click on **Projects**.

3. Select the **New project** drop-down menu and click **New project**.

To create a user project, follow these steps:

1. In GitHub, click on the avatar icon at the top of the page, then select **Your projects**.

2. Select the **New project** drop-down menu and click **New project**.

Let's create a project description or a **README** file:

1. Navigate to your project.

2. At the top right, click to open the menu.

3. In the menu, click **Settings**.

4. Under **Add a description**, type a description in the textbox and click **Save**.

5. Type some content in the textbox to update your project's **README**, under **README**. Then, click **Save**.

Now, let's create a repo so that we can add a project. Go to `https://github.com/` and select **New** to add a new repository.

Exercise – creating an Azure DevOps organization

The first step when starting with Azure DevOps is to create an Azure DevOps organization. An organization is used to connect a group of related projects; this helps us scale up our organization. We can connect to Azure DevOps using a personal Microsoft account or a school or work account. We use this account to connect our organization to our **Azure Active Directory** (**Azure AD**).

Follow these steps to create a new organization in Azure DevOps:

1. Sign in to Azure DevOps and click **New organization**:

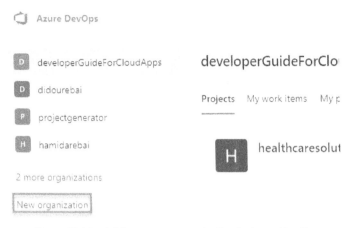

Figure 11.13 – Adding a new organization in Azure DevOps

2. Confirm the name of the Azure DevOps organization and its location. Then, click **Continue** to confirm this:

Figure 11.14 – New organization inputs

3. We can sign in to the organization by going to `https://dev.azure.com/{yourorganization}`.

4. If you are using the free tier with your organization, you can add five users for free, as well as add Azure Pipelines (one concurrent job, up to 30 hours per month for Microsoft-hosted CI/CD and only one self-hosted CI/CD concurrent job), Azure Boards, Azure Repos, and Azure Artifacts.

5. Select **Organization Settings** at the bottom left. Here, we can update the different settings related to an organization:

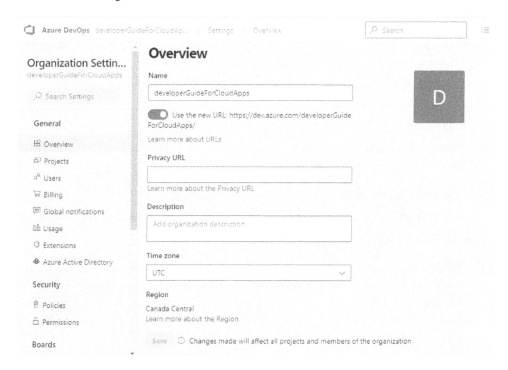

Figure 11.15 – Organization Settings in Azure DevOps

There are multiple tabs under **Organization Settings**:

1. First, we have the **General** tab, which includes the following:

    A. **Overview**: Here, we can update the name of the organization, update the URL by using a privacy URL, add a description, and select a time zone. Once we've done this, we can click **Save**.

    B. **Projects**: Here, we can add a new project by selecting **New project**.

C. **Users**: Here, we can add users and groups of rules. To add a new user, we must select **Add users**. A new dialog window will be displayed where we can select the list of users that will contribute to a selected project in the organization. We can also select an access level – **Basic**, **Stockholder**, or **Visual Studio Subscriber**. We can add a list of projects for specific users. We can also send email invites to different users. DevOps includes group-based licensing for Azure AD groups and Azure DevOps groups. Adding a group rule to assign an access level or extension to a group assigns resources in Azure DevOps to all members of the group. When a user leaves the group, licenses are released back into the pool.

D. **Billing**: Here, we can set up the necessary billing.

E. **Global notifications**: Here, we'll receive notifications about many actions in DevOps, such as when a build is complete or pending deployment approval.

F. **Usage**: This lets us see what's happening in DevOps. We can also filter the information and choose columns by **User** or by **Pipeline**.

G. **Extensions**: We will request a Marketplace extension.

H. **Azure Active Directory**: When accessing DevOps, an email was used to connect to the tenant in Azure because we are using Office 365. Azure DevOps connected to Azure AD.

2. In the **Security** tab, we can add policies and permissions.

3. In the **Boards** tab, we can select the processes that we will use for Agile or Scrum, for example.

4. Next, we have the **Pipelines** tab:

A. **Agent pools**: **DevOps pipeline pools** are extended across the organization so that they can be shared across projects. We don't need to manage individual agents – we can sort them into pools. An agent is a piece of installable software that runs one job at a specific time. With Microsoft-hosted agents, each time we run a pipeline, we get a new VM that is deleted after use.

B. **Settings**: Here, we can enable or disable the different options.

C. **Deployment pools**: These are a logical set of target machines backing a deployment group for a set of projects.

D. **Parallel jobs**: For private jobs, we can purchase parallel jobs for Microsoft-hosted servers. For a public project, we can run 10 parallel jobs for Microsoft-hosted and unlimited parallel jobs for self-hosted servers. A job is a set of one or more build tasks that run sequentially on the same target.

E. **OAuth configuration**: This is an OAuth client configuration that defines the underlying settings that are required to set up service connections in our projects.

5. In the **Repos** tab, under **Repositories**, we can configure different settings, such as allowing users to display their Gravatar. We can also set the default branch name for new repositories.

6. In the **Artifacts** tab, we can check the storage sizes of our artifacts. We can get 2 GB free for our artifacts.

Summary

In this chapter, we defined DevOps as a process enabled by different products to deliver value to end users. Then, we presented the different DevOps tools. We explored Azure DevOps and its different components. We also explored GitHub and created an Azure DevOps organization.

In the next chapter, we will learn how to create a build pipeline in Azure Pipelines, map a manual build step to automated build tasks, publish a build to attribute access for other users, and use templates to build multiple configurations.

Further reading

If you want to learn more about Azure DevOps, go to `https://www.packtpub.com/product/azure-devops-explained/9781800563513`.

12

Creating a Development Pipeline in Azure DevOps

Business requires continuous value creation. When products are delivered to satisfied customers, value is created. When the process silos are completed, value is not created. So, the focus must shift from silos to end-to-end value streams.

The core idea is to create a repeatable, reliable, and incrementally improved process to get software from concept to customer.

Azure Pipelines' purpose is to enable a constant flow of changes to production via an automated software production line. A pipeline divides the software delivery process into different phases. Each phase aims to check the quality of new features from different angles to validate new features and prevent bugs from affecting users. A pipeline also provides feedback to the team. Also, everyone involved in delivering new features can see all the changes.

Azure Pipelines is a full-featured service used to build cross-platform **continuous integration** (**CI**) and **continuous deployment** (**CD**) pipelines. It works with your favorite Git provider and deploys to most major cloud services, including Azure. Azure DevOps offers comprehensive pipeline services.

This chapter will cover the creation of a build pipeline in Azure Pipelines, mapping a manual build step to automated build tasks, publishing them and attributing access for different users, and the use of templates to build multiple configurations.

In this chapter, we're going to cover the following main topics:

- Setting up your Azure DevOps environment
- Creating a build pipeline with Azure Pipelines
- Creating a release pipeline in Azure Pipelines
- Creating a CI/CD pipeline for a GitHub repository using Azure DevOps Starter

Setting up your Azure DevOps environment

In *Chapter 11, Introduction to Azure DevOps and GitHub*, we created an Azure DevOps organization and a new project:

Figure 12.1 – Create a project inside an organization

In this project, we have a source code of our application in Azure Repos, and we will start in the next section by creating a build pipeline using Azure Pipelines.

Creating a build pipeline with Azure Pipelines

CD takes the artifacts produced by a CI pipeline and delivers them to a production-like environment. Understanding how to do this is critical to successful software delivery. Azure DevOps has well-structured YAML syntax, such as stages, jobs, steps, and tasks, which can help you achieve those various tasks.

We will learn some basic terms and parts of Azure Pipelines, which will help you explore how you can use them to deploy better code more efficiently and reliably.

We use Azure Pipelines for the CI and CD solutions because we are able to work with any language and any platform and we can ensure simultaneous deployment to different types of targets. We also have an integration with Azure deployments; we are able to build on any environment (Windows, Linux, or macOS), and we also have an integration with GitHub and can work on open source projects.

Azure Pipelines' core concepts

Before creating a new pipeline, we need to understand the core concepts and parts of Azure Pipelines that will help you further explore how Azure Pipelines can help you deliver better code more efficiently and reliably:

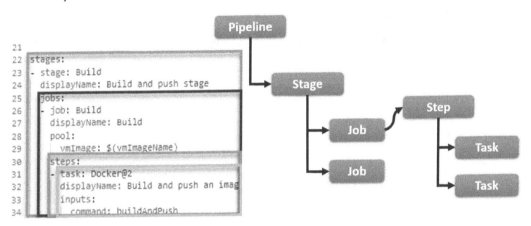

Figure 12.2 – Azure Pipelines core concepts

Pipeline

A pipeline defines the process of CI and delivery of your application. It consists of steps called tasks. It is similar to a script that describes how the test, build, and deploy steps should be performed.

Stage

A stage is the main subdivision of a pipeline. "`Build app`," "`Run integration tests`," and "`Deploy to user acceptance testing`" are some examples of stages.

Job

A build includes one or more jobs. Most jobs run on agents. A job represents an execution boundary for a sequence of steps. All steps run together on the same agent. For example, x86 and x64 can create two configurations. In this case, we have one build and two builds according to the configuration.

Task

Tasks are the building blocks of pipelines. A build pipeline can consist of build and test tasks. A release pipeline is a collection of deployment tasks. Each task executes a specific job in the pipeline.

Creating a build pipeline with Azure Pipelines

We will open an organization in the selected project, and we will click on the pipeline icon to create a build pipeline:

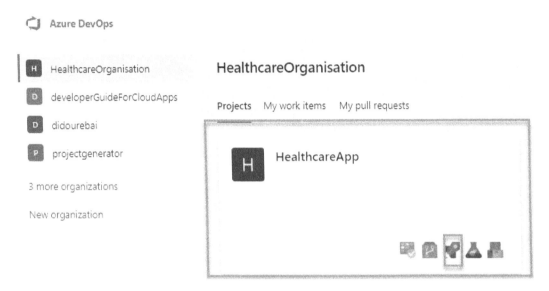

Figure 12.3 – Creating a build pipeline with Azure Pipelines

We will go with a classic pipeline. Select **Create Pipeline** and connect to the source code; in our case, we will select GitHub: `https://github.com/PacktPublishing/A-Developer-s-Guide-to-Cloud-Apps-Using-Microsoft-Azure/tree/main/Chapter12`. Next, we will select a repository from the displayed list of repositories. We can select a private or public repository, and then we need to confirm the authorization on the repository. After that, we will configure the pipeline; we will select **ASP.NET Core (.NET Framework)**.

Azure Pipelines analyzes your repository and recommends ASP.NET Core pipeline templates. After you see your new pipeline, we can explore the YAML file to see the different elements. Once finished, we will select **Save and run**:

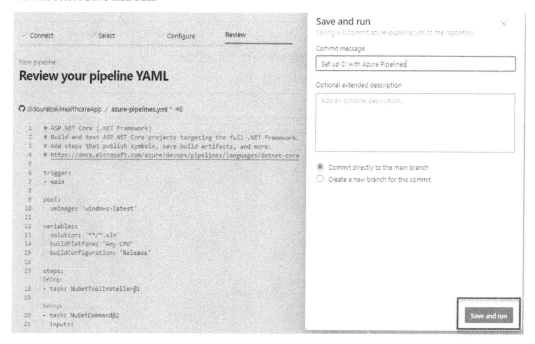

Figure 12.4 – Setting up CI in Azure Pipelines

We will be prompted to commit the new `azure-pipelines.yml` file to our repository. We will select **Save** and try again:

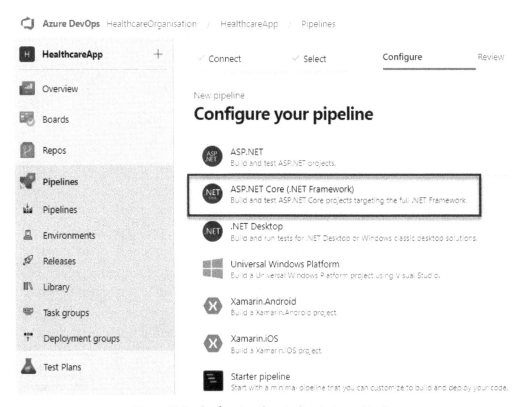

Figure 12.5 – Configuring the pipeline in Azure Pipelines

To display the pipeline in action, we can select build job. We can update the `azure-pipelines.yml` file in the repository according to the application. Let's review the content of the YAML file generated. We can customize the build if required; for example, we can update the pool:

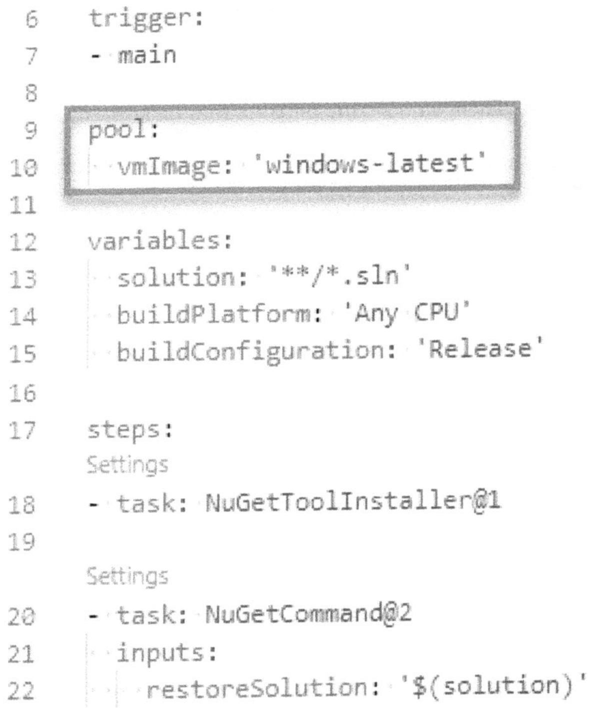

```
 6    trigger:
 7    - main
 8
 9    pool:
10      vmImage: 'windows-latest'
11
12    variables:
13      solution: '**/*.sln'
14      buildPlatform: 'Any CPU'
15      buildConfiguration: 'Release'
16
17    steps:
      Settings
18    - task: NuGetToolInstaller@1
19

      Settings
20    - task: NuGetCommand@2
21      inputs:
22        restoreSolution: '$(solution)'
```

Figure 12.6 – The azure-pipelines.yml file

We can update by using a Linux environment as follows:

```
pool:
  vmImage: "ubuntu-latest"
```

We can also add more scripts or tasks to a pipeline. Plus, we can build and test projects on multiple platforms.

We have created a build pipeline with Azure Pipelines, and in the next section, we will create a release pipeline in Azure.

Creating a release pipeline in Azure Pipelines

Release pipelines are used to automate deployments. This project starts by downloading the artifact files produced by the build pipeline. Then, we split the pipeline into two phases: backup and deployment.

We will navigate to **Releases** under **Pipelines**, and then select **New pipeline**:

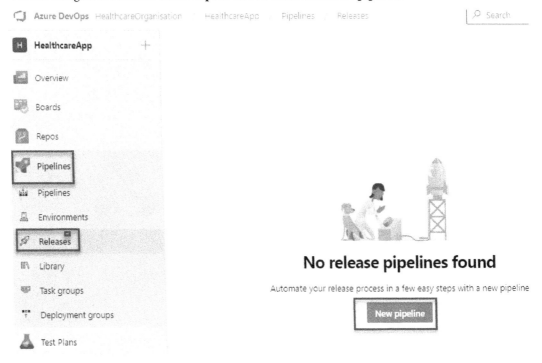

Figure 12.7 – Creating a new release pipeline in Azure DevOps

A new dialog box will be displayed to create a custom template, so click on **Empty job**:

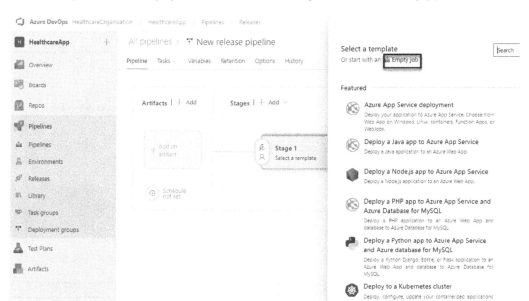

Figure 12.8 – Creating an empty job

An artifact will be selected for deployment.

In this project, **Build** is selected as the source type. Different version control systems can be used if desired. The source alias has been renamed so that it can be used in other tasks:

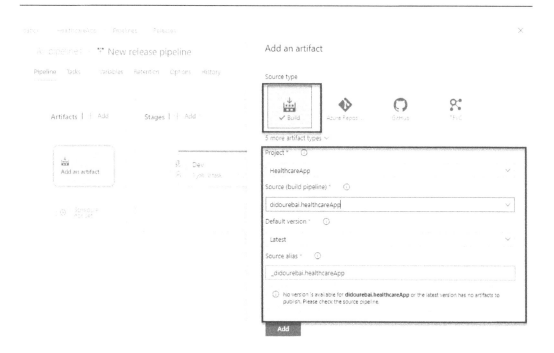

Figure 12.9 – Adding an artifact in Azure DevOps

We will select the name of the project that contains the build pipeline, the name or the ID of the build pipeline that publishes the artifact, the default version, and the source alias.

In **Stages**, we will configure the agent job and save it. Then, we will select **Create release** and a new dialog window will be displayed to select the stages and the artifacts. After that, we will select **Create** to confirm.

Creating a CI/CD pipeline for the GitHub repository using Azure DevOps Starter

In this section, we will create a CI and CD pipeline using the GitHub repository and Azure DevOps Starter.

Azure DevOps Starter is a simplified process for creating a CI and CD pipeline for Azure. We will use existing code from the GitHub repository and a sample template. The sample template offers a dashboard allowing us to monitor code commits, different builds, and deployments through the Azure portal.

We will use DevOps Starter and the Azure portal to create the pipeline. We will configure access to a GitHub repository and select a framework. Then, we will configure Azure DevOps and commit the different changes to GitHub and deploy them to Azure.

Let's create a new DevOps project by following these steps:

1. Sign in to the Azure portal and select **DevOps Starter** in the search box, after selecting **Create**.

2. Select **Bring your own code**, then select **Next**.

    Note that you need to select **Azure DevOps** in the setting up of DevOps Starter with Azure DevOps. To change the settings, we will click on the link that reads **here**. Afterward, in the new dialog window, we select **Azure DevOps**.

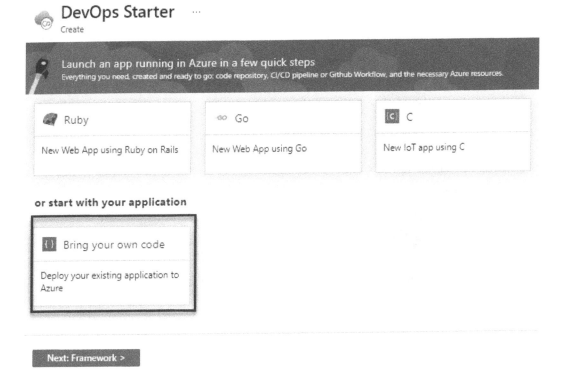

Figure 12.10 – Using DevOps Starter in the Azure portal

3. Select **Next: Framework**, and on the next page, select **GitHub** in **Code repository**:

Figure 12.11 – Selecting GitHub as the code repository

4. Select **Authorize** to allow access to Azure to the GitHub account in order to create a workflow.

5. Select the repository and the branch, then select **Next: Framework** again:

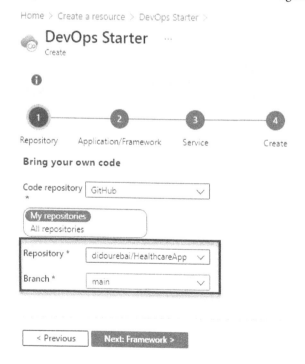

Figure 12.12 – Selecting a repository in Azure DevOps Starter

6. Select the application runtime and the framework and select **YES** for **Is app Dockerized** if your application is already containerized and you have a Dockerfile in the code repository. DevOps Starter will suggest Azure services that can run containerized apps, such as Web App for Containers. Select **Next: Service**.

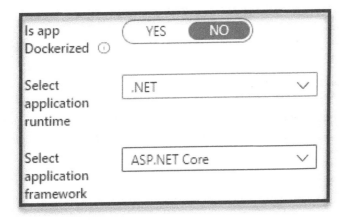

Figure 12.13 – Selecting the application environment

7. Select an Azure service to deploy the application. Here, we can select **Windows Web App** or **Linux Web App**. If we use ASP.NET, we can only select a Windows environment, but for ASP .NET Core, we can select a Linux environment. Select **Next: Create** to finish the configuration.

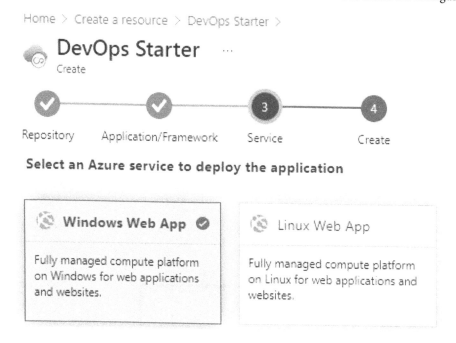

Figure 12.14 – Select an Azure service to deploy the application

8. To finish the configuration to deploy the ASP.NET Core application to the Azure service, we will add a project name. Choose **Azure DevOps Organization**, then **Subscription**. Fill in **Web app name** and **Location**:

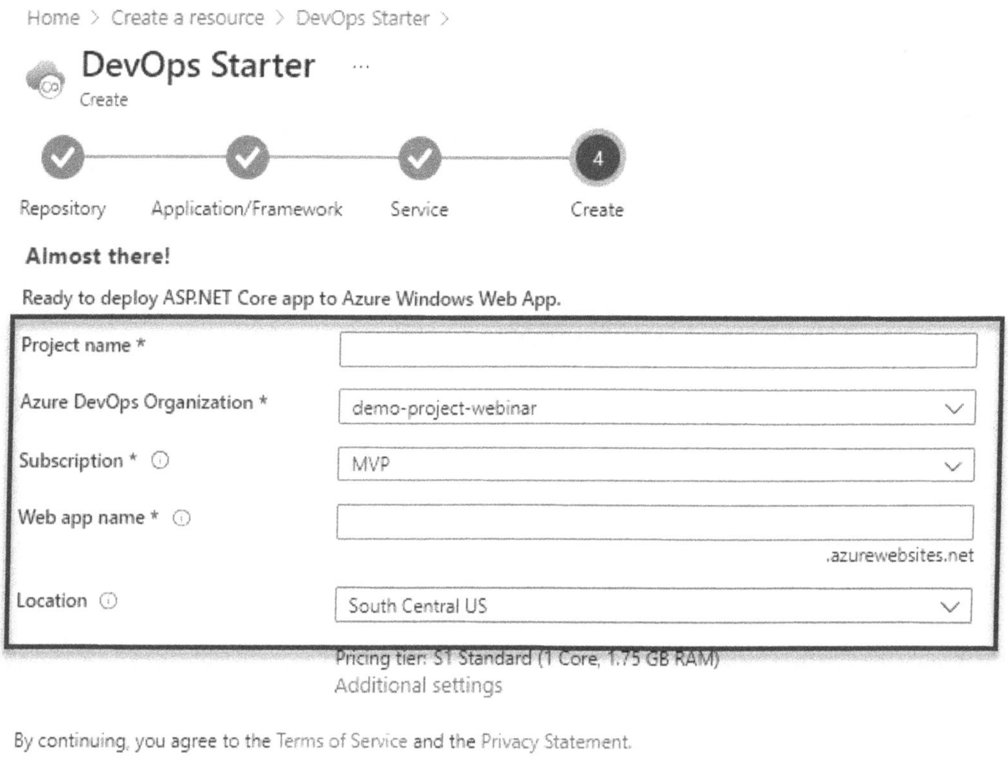

Home > Create a resource > DevOps Starter >

DevOps Starter   ...
Create

Repository   Application/Framework   Service   Create

Almost there!

Ready to deploy ASP.NET Core app to Azure Windows Web App.

Project name *	
Azure DevOps Organization *	demo-project-webinar
Subscription * ⓘ	MVP
Web app name * ⓘ	
	.azurewebsites.net
Location ⓘ	South Central US

Pricing tier: S1 Standard (1 Core, 1.75 GB RAM)
Additional settings

By continuing, you agree to the Terms of Service and the Privacy Statement.

Figure 12.15 – Configuring DevOps Starter to deploy the ASP.NET Core application to Azure

9.   Select **Review + Create**, then select **Create**.

10. After it's created, select **Go to resource** and you will see the CI/CD pipeline.

DevOps Starter automatically configures triggers that deploy code changes to your repository.

Summary

In this chapter, we set up an Azure DevOps environment, created a build pipeline with Azure Pipelines, presented the different core concepts of Azure Pipelines, and created a release pipeline in Azure Pipelines. Finally, we created a CI/CD pipeline for the GitHub repository using Azure DevOps Starter.

This book began with the benefits of moving legacy apps to the cloud and modernizing existing legacy applications using a set of new technologies and approaches. You then learned about the technologies and patterns needed to build cloud-oriented applications. We took a journey through three major services in Azure, namely Azure Container Registry, Azure Container Instances, and Azure Kubernetes Service, which help you build and deploy applications based on microservices. You'll now be able to ensure CI/CD in Azure to fully automate the software delivery process (including the build and release processes).

You should be able to perform application migration assessment and planning, select the right Azure services, and create and implement new cloud-oriented applications using Azure containers and orchestrators.

The second edition of this book will be released to continue our journey in building cloud-oriented applications using serverless and event-driven technologies in Azure, integrating our application with relational or non-relational databases, using database as a service in Azure, and ensuring the CI/CD pipelines of containers on Azure.

Assessments

Chapter 2

- What is the Microsoft Cloud Adoption Framework for Azure? What are the different strategies that we can use for cloud migration?

  Answer: The Microsoft Cloud Adoption Framework for Azure offers guidance on best practices, providing tools, guidance, and narratives to help organizations adopt the cloud and achieve their business outcomes. This framework helps organizations define a robust cloud strategy, plan for successful workload migrations, and ensure complete control over their cloud environment. The tools help organizations to develop technology, business, and people strategies to achieve the best possible business outcomes from their cloud adoption efforts.

  For more details, see the *Understanding the Cloud Adoption Framework* section.

Chapter 6

- To set up a development environment on Windows, what are the necessary tools to install to build and deploy a cloud application? What about for a Linux environment?

  Answers: We need to install Docker Desktop for Windows. For better performance with containers, don't forget to enable `wsl` on your Windows machine, Visual Studio 2022, or Visual Studio Code and Git.

  You can see more details in the *Setting up a development environment on Windows* section. We will install the same tools in a Linux environment, but we will add more packages using command lines. For more details, you can see the *Setting up a development environment on Linux* section.

Index

Symbols

`Packt.com`

Subscribe to our online digital library for full access to over 7,000 books and videos, as well as industry leading tools to help you plan your personal development and advance your career. For more information, please visit our website.

Why subscribe?

- Spend less time learning and more time coding with practical eBooks and Videos from over 4,000 industry professionals

- Improve your learning with Skill Plans built especially for you

- Get a free eBook or video every month

- Fully searchable for easy access to vital information

- Copy and paste, print, and bookmark content

Did you know that Packt offers eBook versions of every book published, with PDF and ePub files available? You can upgrade to the eBook version at `packt.com` and as a print book customer, you are entitled to a discount on the eBook copy. Get in touch with us at `customercare@packtpub.com` for more details.

At `www.packt.com`, you can also read a collection of free technical articles, sign up for a range of free newsletters, and receive exclusive discounts and offers on Packt books and eBooks.

Other Books You May Enjoy

If you enjoyed this book, you may be interested in these other books by Packt:

Infrastructure as Code with Azure Bicep

Yaser Adel Mehraban

ISBN: 978-1-80181-374-7

- Get started with Azure Bicep and install the necessary tools
- Understand the details of how to define resources with Bicep
- Use modules to create templates for different teams in your company
- Optimize templates using expressions, conditions, and loops
- Make customizable templates using parameters, variables, and functions
- Deploy templates locally or from Azure DevOps or GitHub
- Stay on top of your IaC with best practices and industry standards

Azure for Developers - Second Edition

Kamil Mrzygłód

ISBN: 978-1-80324-009-1

- Identify the Azure services that can help you get the results you need
- Implement PaaS components – Azure App Service, Azure SQL, Traffic Manager, CDN, Notification Hubs, and Azure Cognitive Search
- Work with serverless components
- Integrate applications with storage
- Put together messaging components (Event Hubs, Service Bus, and Azure Queue Storage)
- Use Application Insights to create complete monitoring solutions
- Secure solutions using Azure RBAC and manage identities
- Develop fast and scalable cloud applications

Packt is searching for authors like you

If you're interested in becoming an author for Packt, please visit authors.packtpub.com and apply today. We have worked with thousands of developers and tech professionals, just like you, to help them share their insight with the global tech community. You can make a general application, apply for a specific hot topic that we are recruiting an author for, or submit your own idea.

Share your thoughts

Now you've finished *A Developer's Guide to Cloud Apps Using Microsoft Azure*, we'd love to hear your thoughts! Scan the QR code below to go straight to the Amazon review page for this book and share your feedback or leave a review on the site that you purchased it from.

https://packt.link/r/1804614300

Your review is important to us and the tech community and will help us make sure we're delivering excellent quality content.

Download a free PDF copy of this book

Thanks for purchasing this book!

Do you like to read on the go but are unable to carry your print books everywhere?

Is your eBook purchase not compatible with the device of your choice?

Don't worry, now with every Packt book you get a DRM-free PDF version of that book at no cost.

Read anywhere, any place, on any device. Search, copy, and paste code from your favorite technical books directly into your application.

The perks don't stop there, you can get exclusive access to discounts, newsletters, and great free content in your inbox daily

Follow these simple steps to get the benefits:

1. Scan the QR code or visit the link below

https://packt.link/free-ebook/9781804614303

2. Submit your proof of purchase
3. That's it! We'll send your free PDF and other benefits to your email directly

www.ingramcontent.com/pod-product-compliance
Lightning Source LLC
Chambersburg PA
CBHW080633060326

40690CB00021B/4909